Human Ecology and
Climate Change

Cover art: *Presentation of the Sun and Moon*

The cover design, based on the print *Presentation of the Sun and Moon* by Chuna McIntyre, was originally in a screen format. To the Yup'ik (Eskimo) people of Alaska, the Sun and Moon are brother and sister, and are known as the bringers of light. The print shows the people presenting the Sun and Moon to the world to commemorate the incredible gift of light. Mr. McIntyre, a Yup'ik artist, was born in Eek, Alaska near the shores of Kuskokwim Bay on the Bering Sea. He holds a B.A. in Art Studio from Sonoma State University. In addition to being an artist, he is a traditional Yup'ik dancer and storyteller. *Presentation of the Sun and Moon* is reproduced with permission of the artist.

Chapter art: *Young Raven*

The image appearing at the head of each chapter is based on the print *Young Raven* by Kananginak Pootoogook. The original art is a stencil, with black and yellow on a cream background; it is rendered here in black and white. The common raven (Corvus corax) is a ubiquitous creature of the Far North, and is regarded in Northern cultures with a mixture of deference and contempt. The unpredictable raven is described in some stories as clever and witty, and in other stories as greedy and undignified. He is revered as the Creator in some cultures, and provides omens—both good and bad—to people wherever he flies. *Young Raven* is reproduced with permission of the West Baffin Eskimo Co-operative Ltd., Cape Dorset, NWT, Canada.

Human Ecology and Climate Change
People and Resources in the Far North

Edited by

David L. Peterson
University of Washington

Darryll R. Johnson
University of Washington

Routledge
Taylor & Francis Group

LONDON AND NEW YORK

Preface

There are no ordinary moments in the Far North. Both Northerners and southern visitors have visceral impressions of northern landscapes: impressions of biting cold, unrelenting wind, intense summer sun, majestic mountains, crystalline rivers, a shimmering aurora borealis. It is the only region of North America with reasonably intact ecosystems. It is a place where humans are relative newcomers, having lived there only since the last Ice Age.

Recent changes in the atmospheric environment of the Earth may lead to rapid changes in climate during the next century. These changes, a product of southern industrial society, will have the greatest impact on ecosystems at northern latitudes. Dramatic increases in temperature may set off a chain reaction: Terrestrial and aquatic habitats will change, animal populations will be altered, and ultimately human populations may be impacted. People and the land (and water) are inseparable in the Far North. Change the environment and you will change the people.

Scientists have long recognized that climate change could alter northern biological systems, but potential social impacts have received little attention. This is surprising, considering the unique resources and cultures of this region. However, the North has not been able to compete with the political and economic power of the South for attention on the world stage.

In an effort to address this shortcoming, thirty of us gathered at a workshop in October 1993 to discuss how climate change might alter the landscape—biophysical and sociocultural—of the Far North. Biologists, anthropologists, sociologists, and resource managers comprised an enthusiastic group that assessed a wide range of topics. Of all the workshop issues and recommendations, the strongest consensus was on the need to produce this book.

Human Ecology and Climate Change explores how climate change might alter the face of northern North America during the next century. It is an objective look into the future of natural resources and human populations in this region. It is not a comprehensive volume on climate change, northern ecosystems, or northern cultures. Rather, it is an interdisciplinary assessment of these topics. Beyond a gloom-and-doom scenario, we hope the book will offer some societal and managerial approaches for facing an uncertain future.

We are grateful for the efforts of all of the contributors to this volume, many of whom have dedicated a substantial portion of their lives to northern science and culture. We are especially grateful to the workshop participants who provided inspiration and ideas. Technical reviewers provided us with excellent feedback on individual chapters; reviewers include: Harry Bader, Matthew Carroll, Richard Caulfield, Steven Daniels, David Fluharty, Steven Gibbons, William Gregg, Eugene Hunn, Ralph Johnson, Jack Kruse, Conway Leovy, Robert Muth, Gail Osherenko, Gil Pauley, P. Geoffrey Plant, Randall Schalk, Edward Schreiner, Michiyo Shima, Allen Solomon, Joseph Spaeder, Nathan Stephenson, John Walsh, and Andrea Woodward. We thank these colleagues for their helpful comments.

We thank Leila Charbonneau, Christy Parker, June Rugh, and Stephanie Schulz for their assistance with various aspects of technical editing and manuscript preparation. Christy Parker also compiled information for several of the special feature "boxes." Beth Rochefort drafted all figures for final publication. We especially appreciate the professionalism and expertise of Bernadette Capelle and Todd Baldwin of Taylor & Francis, who were consistently supportive of our efforts.

Several individuals and organizations provided essential financial support. Special thanks go to Dr. Pete Comanor (National Biological Service [formerly National Park Service] Global Change Research Program), who had enough confidence in us to provide funding for the workshop and book. The Pacific Northwest Region (Cultural Division, Science and Technology Division) and Alaska Region (Cultural Division, Subsistence Division) of the National Park Service provided generous financial support for the workshop. We also appreciate funding provided by the National Biological Service Division of Cooperative Research, the USDA Forest Service Pacific Global Change Program, and the George Wright Society.

We are particularly grateful for the love and support of Miriam Peterson and Cherry Johnson, who tolerated our all too frequent weekends and evenings of editing. Fortunately, they also know the passion inspired by the North.

It is our sincere hope that this book will stimulate further interest and concern for the northern third of the planet. The book is for students and scholars, resource managers and policymakers, Northerners and Southerners. The Far North needs advocates—for its people and its natural resources—so that it will always be a land without ordinary moments.

David L. Peterson
Darryll R. Johnson

Editors' note: Terms such as *native* and *indigenous people* are used somewhat interchangeably by different authors; the use of these terms has not been modified within individual chapters. The word *Native* is capitalized when it refers to indigenous peoples of North America. The word *North* is capitalized when it refers to northern North America, as in the *Far North*. Arctic is capitalized throughout. Although we have attempted to use consistent terminology, we recognize that readers may have different preferences with respect to wording and usage. We hope that any departures from those preferences will not affect the communication of information and ideas.

Foreword

This is an exciting time to be a sentient organism on the third planet out from the Sun. Almost everywhere on this planet, people are becoming aware that things are not as they used to be. Our world is changing in every aspect. The human component is changing socially, demographically, politically, economically, and spiritually. No less obvious are the changes in what we like to call the Natural World (which, of course, includes all of us) on the surface of the planet: the biosystems, the patterns and flows of water, the distribution of chemicals on land and in the seas and atmosphere, and all of those characteristics that provide habitat for living things.

Of course, the world has never been static. Human societies and biophysicochemical relationships have always been in a state of flux, and it is a delusion to think that there has been a past state of stable normality from which the present situation is an aberration. Nevertheless, there is good reason to believe that the environmental changes that appear to be imminent may be severe enough to have impacts on a planetary scale. These changes seem likely to happen with a rapidity that has never been experienced by our planet. The distinctive feature of this impending change is that much of it is due to the proliferation and activities of a single species: *Homo sapiens*. The response of that species to what might be called "inadvertent self-imposed changes" will to a great degree determine the fate and interrelationship of many other species and all environments.

Regional and global climate is an important component of these changes. Climate, like its local day-to-day manifestation weather, is a dynamic result of energetic and fluid interactions, with wide local variation in energy flow and moisture. On a scale of decades, climate has remarkable stability, although long-term data reflect fluctuation in the planetary heat regime and corresponding changes in terrestrial and aquatic systems. Now, however, the evidence is compelling that changes in atmospheric and Earth surface processes will alter planetary heat flux to produce climatic variability and temperature/moisture rearrangements at a rate not previously experienced by present biosystems.

If human ecology is the study of the relationship of human beings to other forms of life and to their physical and chemical environment, then at the present time it must be a study of rapid change. This book brings together information and insights that address this relationship. To anyone who cares about her or his own future or what is happening to the land in which we live, the presentations are exciting, but at the same time they are sobering and thought provoking. A workshop organized to discuss human-related and institutional aspects of climate change relevant to the management of protected lands in northern North America has triggered perceptive analyses concerning the natural bases for human economic and cultural prosperity and the realities, perspectives, and perceptions of change.

The chapters in this book are focused on northern regions of North America, without defining that area too precisely. This gives the presentations and arguments a coherence with plenty of variety. More than that, the response of the North and its inhabitants to climate change in all its manifestations can serve as an outstanding case study for what may be happening to other regions and to the planet as a whole. If, as the report of the Brundtland Commission so convincingly stated, "The Earth is one, but the World is not," then these chapters on the relationship between changes in this special and beautiful part of the Earth and changes in the world are especially revealing and instructive.

As Sue Ferguson points out, the best evidence from climatology suggests that the combined changes in greenhouse gases, planetary surface albedo, and in upper layers of the ocean will have their greatest relative and absolute effect on the climates of subarctic continental areas. Jonathan Kochmer and Darryll Johnson describe how human societies and cultures adapted to the high-latitude environment whose resources are vulnerable to disruption.

The northern environment and its biological resources vary widely, from the rich fishing grounds of western Alaska to the windswept barrens grazed by musk-ox and migratory caribou. Low annual biological energy flows not only dictate the nature of ecosystems, but determine the likely effects of rapid changes in climate. Patrick Halpin leads us into some interesting aspects of how Arctic landscapes—much more sensitive to climate than the shivering human visitor is sometimes inclined to appreciate—may be affected, and John Hom outlines some fundamental connections between climate and ecosystems in a low-energy setting. While looking at three distinctive groups of Arctic wildlife, Anne Gunn, Kathryn Ono, and Richard Beamish point out not only the marvelous variety of biological systems in the northern environment, but also how certain physiological and behavioral adaptations change rapidly. Northern biosystems have shorter food chains and fewer species compared with lower latitudes. Ecological effects of climate change may be clearer in these systems than in more complex ecosystems elsewhere in the world.

What about the human component? Everywhere in the North, human populations are expanding, and human societies and behavior are undergoing profound change. Much has been made of the material, economic, and cultural changes experienced by resident peoples of the North, or by people from other latitudes who have brought their institutions, technologies, and value systems with them. But surely the greatest change and challenge has been spiritual or philosophical. These changes and challenges have made the northern regions a variety store of social issues: small cultures trying to preserve their fundamental values; entrepreneurs whose vision is limited to economic exploitation; those who see the North as the last place where they can be free of bureaucratic restrictions; governments who see the region as a place to exercise colonial-type control; economies dependent on or enamored with development of living resources or minerals; economists who see the region as a sink for public funds; idealists and common citizens who recognize environmental values that contribute to the spirit of nations; and others who see dismal social conditions with high rates of suicide, alcoholism, and disillusion.

These problems and value conflicts almost certainly outweigh concerns about impending climate change. However, concerns about the effects of changing climate are very real for indigenous cultures, as well as industrialized societies. Steve Langdon, Donald Callaway, and George Wenzel summarize various viewpoints and likely responses of human communities to impending climate change. As with biological systems, the evidence suggests that, at least for a couple of generations, rapid changes in climate may be an additional stress, rather than a blessing, for human populations in the North.

What should humans do in the face of impending rapid climate change? We have suggestions from people who have both theoretical knowledge of and direct experience with these issues in the North. Patrick West analyzes the effect climate change may have on the role of protected areas and their relationship with northern residents. Peter Usher draws on decades of experience with Native and non-Native American ways of managing living resources and looks to the future. John Wiener tackles the thorny problem of management of common property, so difficult for southern cultures but so natural to indigenous northern cultures. Ellen Bielawski, teacher and interpreter of Native American viewpoints, articulates the comprehensive and contextual knowledge base of people who have lived in the North for centuries and developed their own perspective on environmental change. Robert Bosworth points out the basic political and economic facts of subsistence management in the North. Jeanne Schaaf and Martin Price each explore the opportunities and problems of protected areas—Beringian Land Bridge National Preserve and biosphere reserves—as examples and symbols of environmental value and northern spirit and as centers of long-term research and demonstrations of good management. Patrick West and Chip Dennerlein add to this theme with essays on environmental values, resources, and human values. And Stewart Cohen shares some of the interim findings, frustrations, and changing points of view that emerge when an assemblage of research teams, with varied classical science, indigenous, and business backgrounds, address the potential biophysical, cultural, and economic impacts of a rapidly changing climate on a rapidly changing socioeconomic situation in a large northern region.

There is much food for thought here. The editors, David Peterson and Darryll Johnson, leave us some suggestions for further research and better management of the North, its environmental characteristics, and resources. That is what they set out to do. They have succeeded, perhaps beyond their original expectations when they convened the "Human Ecology and Climate Change" workshop. For that, we can all thank them, but they have done much more. They have produced a useful case study about human ecology and climate change that the rest of the world can examine, learn from, or discredit if need be, as similar problems are addressed in other regions. For those of us who live in the North or whose hearts are in the North, they have helped us put our own issues, limited knowledge, and viewpoints in perspective. If a globe can be said to have a corner, this is a wonderful and exciting corner in which to be. But, of course, it is a corner that has its problems too.

E. Fred Roots

Contributors

Richard J. Beamish
Pacific Biological Station
Department of Fisheries and Oceans
Nanaimo, B.C. V9R 5K6
Canada

Ellen Bielawski
Arctic Institute of North America
University of Calgary
2500 University Drive N.W.
Calgary, Alberta T2N 1N4
Canada

Robert Bosworth
Division of Subsistence
Alaska Department of Fish & Game
P.O. Box 25526
Juneau, AK 99802

Donald Callaway
Subsistence Division
National Park Service
2525 Gambell Street
Anchorage, AK 99503-2892

Stewart J. Cohen
Climate and Atmospheric Research
Directorate
Environment Canada
4905 Dufferin Street
Downsview, Ontario M3H 5P4
Canada

Chip Dennerlein
National Parks and Conservation
Association
329 F. Street #208
Anchorage, AK 99516

Sue A. Ferguson
USDA Forest Service
Pacific Northwest Research Station
4043 Roosevelt Way N.E.
Seattle, WA 98105

Anne Gunn
Department of Renewable Resources
Government of the Northwest
Territories
Yellowknife, N.W.T. X1A 3S8
Canada

Patrick N. Halpin
Department of Environmental
Sciences
Clark Hall
University of Virginia
Charlottesville, VA 22903

John L. Hom
USDA Forest Service
5 Radnor Corporate Center
P.O. Box 6775
100 Matsonford Rd., Suite 200
Radnor, PA 19087-4585

Darryll R. Johnson
Cooperative Park Studies Unit
Box 352100
University of Washington
Seattle, WA 98195

Apangulak Charlie Kairaiuak
1574 Wintergreen Street
Anchorage, AK 99508

Jonathan P. Kochmer
University Extension
University of Washington
1401 N. 40th Street
Seattle, WA 98103

Steve J. Langdon
Department of Anthropology
3211 Providence Drive
University of Alaska
Anchorage, AK 99508

Kathryn A. Ono
Department of Life Sciences
University of New England
Hills Beach Road
Biddeford, ME 04005

Christy M. Parker
Cooperative Park Studies Unit
Box 352100
University of Washington
Seattle, WA 98195

David L. Peterson
Cooperative Park Studies Unit
Box 352100
University of Washington
Seattle, WA 98195

Martin F. Price
Environmental Change Unit
University of Oxford
1a Mansfield Road
Oxford OX1 3TB
Great Britain

E. Fred Roots
UNESCO-MAB Northern Sciences
Network
R.R. #3
Wakefield, Quebec J0X 3G0
Canada

Jeanne Schaaf
Division of Cultural Resources
National Park Service
2525 Gambell Street
Anchorage, AK 99503-2892

Peter J. Usher
P. J. Usher Consulting Services
Box 4815, Station E
Ottawa, Ontario K1S 5H9
Canada

George W. Wenzel
Department of Geography
McGill University
Montreal, Quebec H3A 2K6
Canada

Patrick C. West
School of Natural Resources &
Environment
Dana Building
University of Michigan
Ann Arbor, MI 48109-1115

John Wiener
Environment and Behavior Program
Institute of Behavioral Science
University of Colorado
Boulder, CO 80309-0468

List of Figures and Tables

Many informative tables, maps, illustrations, and photographs are used in this book. The following provides a guide to these features:

FIGURES

TABLES

PART I

Climate and Human Populations— A Dynamic Balance

Human technological development has evolved faster than its ability to deal with its own waste products. Degradation of the atmosphere by modern industrial society has left us with potential problems of global proportions. If we are to address these problems, we need to clearly define them prior to taking any action. We must know where we are—and a little bit about where we have been—before we face an uncertain future.

Describing the climatic systems of northern North America is a daunting task. Offhand critiques of the daily weather forecast tend to ignore the remarkable complexity of the atmosphere. Fortunately, this section sheds some light on exactly how northern climates function and how they might change in the next century. The predictions are ominous. If Earth surface temperatures increase due to an accumulation of greenhouse gases, the greatest increases will occur at northern latitudes. The complexity (and uncertainty) of northern climate is complicated even further by potential feedbacks from hydrologic and biological systems.

It is equally challenging to describe the structure of northern social systems. Demographers and economists have never been very successful at predicting long-term trends despite using the best available information. One thing we know for certain is that populations in northern North America are increasing. We also know that the cultures and lifeways of indigenous people are under considerable stress, and that the path toward acculturation has had tremendous social and economic costs. The social and demographic trends discussed here indicate that some segments of northern society will be quite vulnerable to changes in natural resource availability. Communities where social stress is already high cannot tolerate much more.

This section provides some background for assessing the potential impacts of climate change in the North. Rapid changes in climate and natural resources would have a wide range of impacts on people who depend on those resources. And it is not difficult to imagine that a warmer climate would facilitate the transfer of many economic activities from the South. A wave of immigration to the North would change northern landscapes and cultures forever.

1

Human ecology and climate change at northern latitudes

David L. Peterson and Darryll R. Johnson

Imagine you are a hunter living in Alaska 14,000 years ago. Your world is filled with large and frightening creatures such as the sabertooth (*Smilodon fatalis*), dire wolf (*Canis dirus*), mastodon (*Mammut americanum*), and wooly mammoth (*Mammuthus primigenius*). While you have the tools and knowledge to hunt many of these animals, you are also potential prey for some of the larger carnivores. The paleontological record contains a fascinating account of the number and diversity of large mammals that roamed the landscapes of North America at the end of the last Ice Age.

Then, between 12,000 and 10,000 B.P. (years before present)—a veritable blink of the eye in geological terms—huge numbers of mammals disappeared from the face of the earth. At least 40 large mammal species in North America became extinct (Kurtén and Anderson 1980; Pielou 1991). Some 200 mammalian genera became extinct worldwide (Behrensmeyer et al. 1992). This is a remarkable change in biological diversity over a relatively short time period. While there are other examples of massive extinctions (e.g., dinosaurs at the end of the Cretaceous period), the Pleistocene extinctions are particularly intriguing because they were so specific to the so-called mammalian "megafauna." There is no consensus on the cause of the extinctions.

There are a few things we know for sure about the post-Ice Age world. One is that average earth surface temperature increased 5°C as glacial ice sheets and continental glaciers receded (Chapter 2). This resulted in large-scale changes in the distribution of plant species and animal habitat at northern latitudes. Tundra was transformed into forest, and new areas of tundra and wetlands were created along the margins of the retreating ice sheet. Changes in the distribution of habitat during this period may have contributed to the massive extinctions (King and Saunders 1984).

Another thing we know about the post-Ice Age world is that human populations were expanding in North America. The ancestors of the Clovis people, who reached North America from Beringia ca. 14,000–12,000 B.P., dispersed rapidly through Alaska, northern Canada, and southerly locations (West 1983; Kunz and Reanier 1994), coincident with the mammal extinctions. These early inhabitants were prolific hunters, and it is possible that many species were vulnerable to this tool-using human predator (Pielou 1991). The notion that humans facilitated the demise of Pleistocene mammalian fauna has been termed the "prehistoric overkill (or blitzkrieg) hypothesis" (Martin 1984; Behrensmeyer et al. 1992).

So what really caused the post-Pleistocene extinctions? We may never know all the details, but the extinctions were no doubt the result of changes in both environmental conditions and human populations (Burney 1993). Regardless of the relative importance of these factors, prehistoric evidence clearly illustrates linkages among climate, natural resources, and humans that have existed for millennia.

Climatic variation and human populations

The paleoecological literature tells a story of dramatic change in the distribution of North American vegetation during the Holocene era (since 12,000 years B.P.). A true long-term equilibrium between climate and ecosystems probably never exists. Even submillennial variations in climate such as during the Medieval Optimum (ca. 1000–1200 A.D.) and the Little Ice Age (ca. 1600–1850 A.D.) can have significant impacts on fire regimes and vegetation distribution (Delcourt and Delcourt 1991). Similarly, true equilibrium between social and ecological systems never exists because the systems and their interactions change continuously.

The impacts of climate on human populations are perhaps better documented for the temperate and arid regions of North America than for the Far North. A prime example of the effects of environmental change on human populations occurred in what is now the southwestern United States. A prolonged drought at the end of the thirteenth century forced the Anasazi people to abandon their extensive cliff dwellings in Canyon de Chelley (Arizona) and Mesa Verde (Colorado), when agriculture was no longer viable (Watson 1953). The archeological record contains many examples in the American Southwest of human communities abandoning areas they had occupied for hundreds of years due to changing climate and environmental conditions. A more recent example of climate-related social impacts occurred in the 1930s, when prolonged droughts caused widespread agricultural failures and soil erosion in the dust bowls of the central United States. As recorded in Woody Guthrie's songs and John Steinbeck's novels, thousands of people abandoned their homes and sought better economic opportunities by moving to the far western states.

There are recent examples from other parts of the world as well. Droughts during the 1980s, combined with excessive use of native vegetation, have resulted in increasing desertification of the Sahel region of Africa and decreasing land area for human habitation. Particularly high amplitude of the El Niño Southern Oscillation phenomenon in the early 1980s caused a drastic reduction in fish populations and food supply in some areas of western South America (Diaz and Markgraf 1993). Variation in climate can impact human cultures that subsist directly on natural resources (hunting, fishing, gathering) (Chapters 9 and 11), as well as those that manipulate the landscape for agriculture.

Less known, but equally dramatic, are the impacts of climatic variation on indigenous peoples of the Far North (Chapter 11). Beginning about 1000 A.D., the Arctic experienced a warm period with summer temperatures approximately 2°C higher than they are today. This so-called Neo-Atlantic Optimum caused a decrease in the duration and distribution of summer sea ice in the eastern Canadian Arctic, extending maritime habitat to a variety of migratory whales (Cetacea) and seals (Pinnipedia). Of all the changes in the marine mammal populations during this period, none was more important to the cultural history of the region than increased numbers of the bowhead whale (*Balaena mysticetus*).

Prior to 1000 A.D., the eastern Canadian Arctic was dominated by the Dorset culture. Subsistence activities consisted primarily of hunting ringed seal (*Phoca hispida*) in winter, hunting migratory caribou (*Rangifer tarandus*), and fishing for Arctic char (*Salvelinus alpinus*) in summer (Chapter 11). As warmer temperatures and more open waters of the eastern Arctic facilitated the movement of bowheads, migrants from Alaska increasingly hunted whales in the area. The Dorset culture was rapidly replaced by the Thule culture, which brought whale-hunting technology and a variety of technological innovations and cultural traditions. The Thule people are the forerunners of the Inuit people who live in this region today (Chapter 11). This and other examples of climate-induced changes in social and cultural systems of northern North America (e.g., McGhee 1992) are rarely included in recent discussions of the social impacts of climate change.

Climate change, natural resources, and humans

Average earth surface temperature may increase 2°–5°C during the next century as a result of increased concentrations of greenhouse gases in the atmosphere, with even greater temperature increases (up to 10°C) in the Arctic regions (Chapter 2). Some ecologists have suggested that climate during the warm Altithermal period (9,000–5,000 B.P.) was an approximate analog for the predicted "greenhouse world" of the next century. The validity of this analogy is uncertain, but we do know that the beginning and end of this period were marked by dramatic changes in vegetation distribution, especially at northern latitudes (Bliss and Richards 1982). Ecosystems are extremely dynamic; rapid changes in future climate may or may not cause widespread extinctions, but there will almost certainly be large-scale changes in the distribution and abundance of plant and animal species (Chapters 4–8).

Moderate changes in *mean* temperature are probably not as much of a concern as changes in the *variance* of the climatic system. Greater variation in climate could result in different frequencies of drought and extreme cold. Greater amplitude of the El Niño Southern Oscillation could have tremendous impacts on this variation (Diaz and Markgraf 1993). Changes in the occurrence of extreme climatic conditions could affect the frequency of disturbance events, such as windstorms and fires, which have large-scale impacts on terrestrial ecosystems (Chapter 5).

Humans have dramatically altered North American landscapes during the past several thousand years, particularly during the last century. Biological diversity in temperate regions has been reduced through resource exploitation and habitat loss, resulting in smaller populations of many terrestrial and aquatic taxa. If a local habitat becomes unfavorable for an animal due to climate change, human-caused alterations (such as agriculture and urban areas) may interrupt migration routes to another favorable habitat. Biological systems and taxa experiencing stress due to reduced habitat could be especially vulnerable to additional stress from climate change. Humans are, of course, the *cause* of increased greenhouse gases in the atmosphere and provide direct feedbacks between social and biological systems (Figure 1.1).

Much of the recent literature on climate change portrays a "gloom and doom" future for biological and social systems. Whether future changes would be detrimental or beneficial is subjective. Climatic patterns that cause crop failures in one region may increase agricultural production in another. The same may be true of forest growth (Chapter 5), fish populations (Chapter 8), and the movement of caribou herds (Chapter 6). As biological systems adjust to climate change with a redistribution of resources, social systems and economies may also respond with a period of reorganization (Chapters 9, 11, 14, and 16). The rate at which climate and human demographic and sociocultural patterns (Chapter 3) change will determine the extent of disequilibrium affecting the climatic system, biological systems, and social systems.

Resource management and human populations in the Far North

The northern latitudes of North America contain a large proportion of lands that remain relatively unaltered by humans. Small human populations and harsh climate have reduced the impact of humans on the integrity of natural ecosystems. There are also substantial areas with protected status[1] at northern latitudes (Fig-

---

[1] The IUCN (World Conservation Union) Commission on National Parks and Protected Areas defines a "protected area" as any area of land that has legal measures limiting human use of the plants and animals within that area (IUCN 1994). Six management categories of protected areas are defined within the IUCN Guidelines for Protected Areas: (1) strict protection (e.g., nature reserve/wilderness area), (2) ecosystem conservation and recreation (e.g., national park), (3) conservation of natural features (e.g., natural monument), (4) conservation through active management (e.g., habitat/species management area), (5) landscape/seascape conservation and recreation (e.g., protected landscape/seascape), and (6) sustainable use of natural ecosystems (e.g, managed resource protected area).

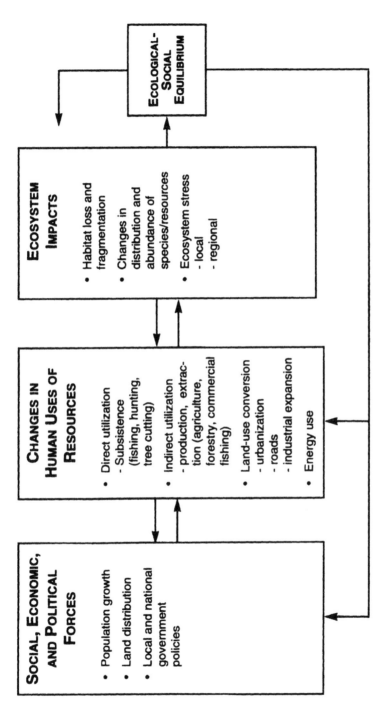

Figure 1.1. Conceptual model of ecological-social interactions relevant to the impacts of potential climate change. Adapted from Machlis (1992), figure 1. Reprinted with kind permission from Elsevier Science Ltd, The Boulevard, Langford Lane, Killington OX5 1GB, United Kingdom.

ures 1.2 & 1.3). For example, nearly 70% of all United States national park lands (18.4 million hectares) are in Alaska. Huge expanses of sparsely populated boreal forest, tundra, lakes, and rivers exist throughout many of the provinces and the Yukon and Northwest Territories of Canada.

Rural Northerners, especially indigenous people, live in remote communities with subsistence-based economies that depend on wildland resources for personal consumption and social purposes. In these economies, substantial wild fish, meat, and plant harvests are supplemented by participation in the cash economy, resulting in a mixed subsistence-cash system (Chapters 9–11). Such social systems are characterized by networks of exchange and sharing of subsistence products that reinforce individual identity and provide social bonds linking individuals to family and community.

Northern rural communities in and near protected areas often face a complex maze of political institutions involved in resource management (Chapter 17). In some Alaskan villages, there may be six federal agencies, the Alaska Department of Fish and Game, and numerous local and tribal government officials involved in the management of lands used for subsistence purposes. Each of these groups operates with different mandates, legal foundations, and political constituencies.

Protected areas throughout the Far North have been designated for a variety of reasons, including the protection of natural features, commodities, and human values. This includes national parks, national forests, wilderness areas, and wildlife refuges in the United States and national parks, provincial parks, and Crown lands (with various status) in Canada. Social and political forces have had a significant impact on when and where protected areas are designated. More recently there has been greater emphasis on preserving ecosystems in a more systematic way through the designation of reserves that represent specific biological and cultural characteristics (Chapter 18).

Establishment and management of North American protected areas often tend to exclude or ignore human interactions (past and present), with the exception of nonconsumptive recreational activities. This perspective contrasts with attitudes and approaches in other regions of the world, including some developing countries, in which local populations are integrated into the establishment and management of protected areas. A common approach is to designate a core protected zone with peripheral areas for increasing levels of resource use at increasing distance from the core (Chapter 18). Although protected areas are often perceived as refuges, they are also an important source of biological and cultural diversity, especially with respect to the impacts of climate change (Peterson et al. 1993).

Rural populations in the North are, like the biota, a valuable resource in themselves. Indigenous and other peoples who live a subsistence or mixed subsistence-cash lifestyle use relatively little energy compared to Southern urban populations. They generally use resources without severely depleting them. Conversely, com-

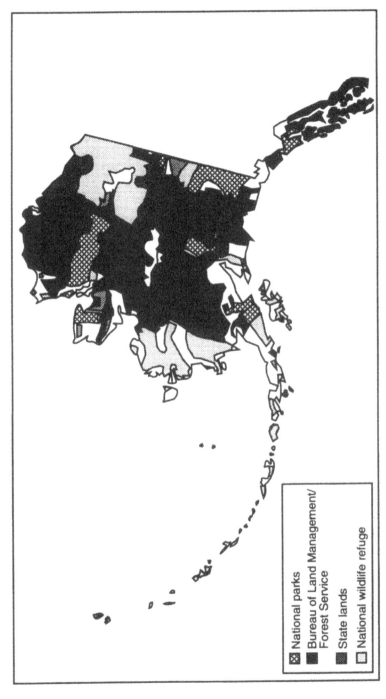

Figure 1.2. Alaska contains approximately 91 million hectares of land in protected areas (IUCN 1994), grouped here by general category.

Legend:
- National parks
- Bureau of Land Management/Forest Service
- State lands
- National wildlife refuge

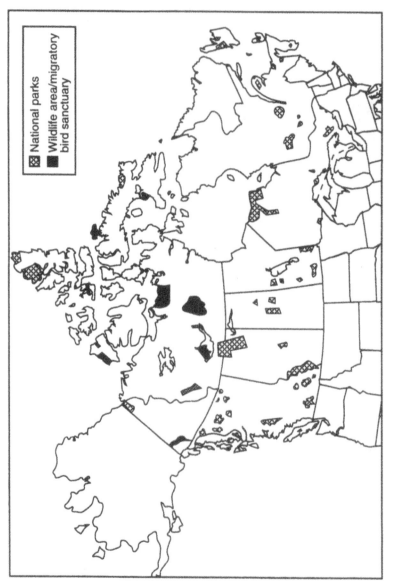

Figure 1.3. Canada contains approximately 83 million hectares of land in protected status (IUNC 1994), grouped here by general category. In addition, provincial and federal Crown lands comprise 48% (provinces) and 39% (Yukon and Northwest Territories) of Canada, respectively. Crown Lands include some of the protected areas indicated on the map and also include lands with various protected status (i.e., long-term management of forests and other resources) that are difficult to identify and classify geographically.

National parks

Wildlife area/migratory bird sanctuary

mercially valued resources such as fisheries can be managed responsibly and harvested on a sustainable basis without damaging subsistence harvests.

Climate change and the biosocial environment

Dramatic changes in the distribution and abundance of natural resources may—or
may not—occur during the next century. There is a high degree of uncertainty
because we have only a rudimentary capability of modeling the complex atmospheric and biological systems of the Earth. Human demographic and sociocultural patterns are complex and equally difficult to model. Linking biological
impacts to social impacts adds yet another layer of complexity to assessments of
climate change.

Temperature may increase at the global scale, but most human populations
perceive the environment at the local or regional scale. Gradual temperature
change may not seem very important to a subsistence hunter in the Northwest
Territories compared to the urgency of supplying food for his village. Climate
change may be an equally small consideration for an office worker in Yellowknife
(Northwest Territories). Increased flooding of northern villages by rivers during
the spring may be a tangible climate change impact that would restrict subsistence hunting routes and alter wildlife distribution. Likewise, flooding might
impair local transportation routes used by the office worker. However, climate
change impacts that are merely an inconvenience to the office worker may
threaten the very livelihood and culture of the northern subsistence hunter.

Rural Northerners have a good understanding of natural resources through
observation and deduction (Chapter 15). Most Native communities, however,
have less understanding and little tolerance for the regulations of agencies and
institutions that govern their lands and activities. How will northern communities
deal with possible changes in wildlife and fish populations due to climate change,
when there are already serious conflicts over the use of resources?

Some creative methodologies are needed to incorporate the complexity of
the biological and social environment with an assessment of climate change. Fortunately, there are several approaches that will assist in this assessment. Comanagement (Chapter 13) and common property management (Chapter 14) are
concepts that deserve consideration as institutional solutions to resource management across political and sociocultural boundaries. Conflict management is
essential for the implementation of these or any long-term institutional frame-
works intended to coordinate the use of resources among local populations and
agencies (Chapter 12).

If any scientific or resource management effort in the Far North is to have
long-term success, one thing is certain: local populations and stakeholders must
be integrated in the effort (Chapters 16 and 21). This has been demonstrated by
the successful implementation of biosphere reserves in many regions of the world

(Chapter 18). Integration of a wide range of stakeholders has also been achieved at the regional level, with special emphasis on climate change, for the Mackenzie River Basin in Canada (Chapter 21). These efforts may not result in consensus policies and actions. In fact, different perspectives (Chapters 19 and 20) can and should be integrated into assessments and management plans as much as possible. Only by including a diversity of viewpoints will stakeholders have true ownership of the final policy or action.

Alaska and northern Canada contain the last large, reasonably intact, natural ecosystems in North America. These ecosystems play a critical role in climate change assessment at the global scale, because they store a large proportion of the Earth's carbon. The Far North also contains the primary human populations on the continent that maintain a strong connection to the land through subsistence on natural resources. The subsistence practices and cultures of these indigenous people have endured over thousands of years, but the connection of these people to the land has been threatened by dominant Euro-American culture and institutions over the past century. Will climate change be the stress that breaks the link between the land and the people? Hopefully, no. We must consider northern climate change aggressively for the sake of the land *and* the people. This will enable us to work toward a biosocial environment in which humans—as a part of ecosystems—can coexist with the natural world in perpetuity.

References

Behrensmeyer, A. K., J. D. Damuth, W. A. DiMichele, R. Potts, H. D. Sues, and S. L. Wing. 1992. *Terrestrial ecosystems through time: Evolutionary paleoecology of terrestrial plants and animals.* Chicago: University of Chicago Press.

Bliss, L. C., and J. H. Richards. 1982. Present-day arctic vegetation and ecosystems as a predictive tool for the Arctic-Steppe Mammoth biome. In *Paleoecology of Beringia*, edited by D. M. N. Hopkins, J. V. Matthews, C. E. Schweger, and S. B. Young, 241–257. London: Academic Press.

Burney, D. A. 1993. Recent animal extinctions: Recipes for disaster. *American Scientist* 81:530–541.

Delcourt, H. R., and P. A. Delcourt. 1991. *Quaternary ecology: A paleoecological perspective.* New York: Chapman and Hall.

Diaz, H. F., and V. Markgraf. 1993. *El Niño: Historical and paleoclimatic aspects of the Southern Oscillation.* New York: Cambridge University Press.

IUCN (World Conservation Union). 1994. *1993 list of United Nations national parks and protected areas.* Gland, Switzerland: World Conservation Monitoring Centre.

King, J. E., and J. J. Saunders. 1984. Environmental insularity and the extinction of the American mastodont. In *Quaternary extinctions*, edited by P. S. Martin and R. G. Klein, 315–319. Tucson: University of Arizona Press.

Kunz, M. L., and R. E. Reanier. 1994. Paleoindians in Beringia: Evidence from Arctic Alaska. *Science* 263:660–662.

Kurtén, B., and E. Anderson. 1980. *Pleistocene mammals of North America*. New York: Columbia University Press.

Machlis, G. E. 1992. The contribution of sociology to biodiversity research and management. *Biological Conservation* 62:161–170.

Martin, P. S. 1984. Prehistoric overkill: The global model. In *Quaternary extinctions*, edited by P. S. Martin and R. G. Klein, 354–403. Tucson: University of Arizona Press.

McGhee, R. 1992. Archaeological evidence for climatic change during the last 5,000 years. In *Climate and history: Studies in past climates and their impact on man*, edited by T. M. L. Wigley, M. J. Ingram, and G. Farmer, 162–179. Cambridge: Cambridge University Press.

Peterson, D. L., A. Woodward, E. G. Schreiner, and R. D. Hammer. 1993. Global environmental change in mountain protected areas: Consequences for management. In *Parks, peaks, and people,* edited by L. S. Hamilton, D. P. Bauer, and H. F. Takeuchi, 29–36. Honolulu, HI: East-West Center Program on the Environment.

Pielou, E. C. 1991. *After the Ice Age*. Chicago: University of Chicago Press.

Watson, D. 1953. *Indians and the Mesa Verde*. Washington, DC: National Park Service.

West, F. H. 1983. The antiquity of man. In *Late Quaternary environments of the United States, vol. 1, the Late Pleistocene*, edited by S. C. Porter, 364–382. Minneapolis: University of Minnesota Press.

2

Potential climate change in northern North America

Sue A. Ferguson

A frozen wasteland of endless winter does not sound very susceptible to change. Despite this common description of Arctic landscapes, northern North America is quite vulnerable and can change dramatically. For example, in late summer the extent of Arctic sea ice can fluctuate by as much as 1 million square kilometers for a 1°C change in average summer temperature (Meier et al. 1985). When globally averaged temperatures change, most of the change is manifested over land areas of northern latitude. At these high latitudes, the increase in winter-averaged temperatures due to global warming may be three to six times more than globally averaged temperatures (IPCC 1990).

Climate change, whether natural or human-caused, is imminent. Anticipating the direction and magnitude of change requires knowledge of natural climatic fluctuations as well as the impacts of human activity. Of all the recent climate change issues, the potential for global warming due to enhanced greenhouse gases has received the most attention. The concentration of atmospheric carbon dioxide (CO_2), a significant greenhouse gas (any gas or vapor that is relatively transparent to incoming short-wave solar radiation but effectively absorbs outgoing long-wave radiation), has clearly been increasing since the Industrial Revolution.

General circulation models (GCMs) are used to test the sensitivity of the Earth's environment to increases in CO_2 concentration, or its greenhouse gas equivalent. Early GCMs (e.g., Schlesinger and Zhao 1989) indicated that doubling CO_2 concentrations could increase average surface temperatures from 1.9° to 5.2°C over present temperatures. The Intergovernmental Panel on Climate Change (IPCC) suggests that 2.5°C is a "best guess" for anticipated global warming (IPCC 1990). If atmospheric CO_2 concentration continues to increase at its present rate, this magnitude of warming may be possible within the next 50 to 100 years.

Transient (time dependent) global models that couple ocean and atmospheric dynamics (e.g., Manabe et al. 1991) are less sensitive than GCMs to the effects of increasing CO_2. If CO_2 concentrations increase at their present rate, coupled GCMs (CGCMs) estimate that globally averaged surface temperatures could increase between 0.7° to 1.5°C over present temperatures by the year 2030. This rate of warming would be greater than at any time during the past 10,000 years (IPCC 1990, 1992).

Because GCMs simulate the global atmospheric environment at coarse spatial and temporal scales, they simulate regional climate poorly (Giorgi and Mearns 1991). The models are also deficient in simulating Arctic inversions, sea-level pressure patterns, and storm activity, because they do not include realistic ice dynamics and thermodynamics (Walsh and Crane 1992).

The challenge in understanding regional response to climate change is to downscale the large climatic patterns simulated in the GCMs to spatial and temporal resolutions that are meaningful. Giorgi and Mearns (1991) describe three basic methods to simulate regional climate change: (1) empirical, from historical or paleoclimatic analogs; (2) semi-empirical, in which GCMs describe large-scale patterns and empirical techniques are used to describe regional patterns; and (3) modeling, in which smaller-scale physical models are nested within the GCMs. Most efforts to describe potential change in Arctic climate have been through empirical and semi-empirical approaches (e.g., McBeath 1984; Hare 1988; Maxwell 1992; Weller 1993).

Environmental controls on high-latitude climate

Greenhouse effect

If the Earth had no atmosphere, its average surface temperature would be about -19°C. Because of a natural greenhouse effect, the average surface temperature of the Earth is a comfortable 14°C. Greenhouse gases, common components of Earth's atmosphere, are nearly transparent to incoming short-wave radiation from the sun. Therefore, the sun's energy can easily penetrate the atmosphere to warm the Earth's surface, just as sunlight penetrates through the glass of a greenhouse. These greenhouse gases are nearly opaque to outgoing long-wave radiation from the Earth, causing heat to be trapped within the "greenhouse" near the Earth's surface (Figure 2.1).

Because of the Earth's gravitational field, most of the atmosphere's mass and water vapor are held within a few kilometers of the Earth's surface. This area, called the troposphere, is where the strongest greenhouse effect occurs. The tropopause, a boundary at the top of the troposphere, is only 16–18 kilometers above the equator and only 8–10 kilometers above the poles. The troposphere is only a thin veil of atmosphere around the Earth's surface.

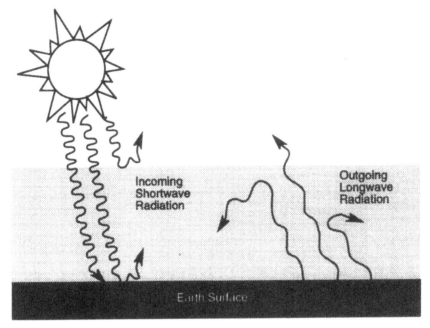

Figure 2.1. Schematic diagram of the greenhouse effect. The Earth's lower atmosphere is nearly transparent to incoming short-wave radiation and nearly opaque to outgoing long-wave radiation. This helps to trap heat near the Earth's surface.

Increases in human activity may be enhancing Earth's natural greenhouse effect. For example, input of CO_2 to the atmosphere from fossil fuel combustion has increased nearly exponentially since 1860. During that time, a similar amount of CO_2 has been released to the atmosphere from deforestation, including burning, biomass decay, oxidation of wood products, oxidation of soil carbon, and regrowth of trees. Concentrations of other greenhouse gases (e.g., methane, nitrous oxide, chlorofluorocarbons) have also increased, although atmospheric CO_2 concentration is generally used to demonstrate effective greenhouse gas contribution. Since 1860, CO_2 concentrations have increased at about 0.5% per year (IPCC 1990). At this rate, we can expect double the pre-Industrial Revolution CO_2 concentration within the next 50 to 100 years.

Water vapor also is a significant greenhouse gas, and a warmer atmosphere may coincide with an increase in atmospheric moisture (IPCC 1990, 1992). This may result in greater cloud cover over much of the globe and more precipitation in some regions. Increased atmospheric water vapor can affect the annual range in daily temperatures. For example, the mean annual range of daily temperatures in Alaska has decreased by 2.4°C since 1950, the likely result of increased cloud cover (Karl et al. 1993). Change in the daily range of temperature could have a significant effect on energy flux between the atmosphere and vegetation or between snow and ice.

Solar radiation

The daily rotational axis of Earth currently tilts about 23.5 degrees. As the Earth makes its yearly orbit, it is tilted away from the sun during Northern Hemisphere winter and toward the sun during Northern Hemisphere summer. A slight wobble in the Earth's spin and variations in its tilt axis and orbital pattern can cause variations in the amount and distribution of solar radiation that reaches the Earth's surface. These variations contribute to glacial–interglacial climatic cycles; globally averaged surface temperatures have fluctuated by as much as 5° to 7°C between major ice ages (Figure 2.2a). Although temperature fluctuations during the most recent interglacial period have typically been less than two degrees (Figures 2.2b & 2.2c), climatic fluctuations during previous interglacial periods are thought to have been much greater. This may indicate that the climate is quite vulnerable to human-induced atmospheric disturbance.

Above the Arctic Circle (66.5°N) there is virtually no direct solar radiation for several weeks to months each winter. With little incoming radiation, a net loss of heat from the surface every winter should continue at these darkened high latitudes, even if globally averaged temperatures increase. Therefore, water in the soil, ocean, rivers, lakes, and clouds should continue to freeze for at least the next 100 winters. This means that some amount of permafrost, sea ice, lake and river ice, and snow cover will remain. The magnitude and duration of winter freezing, however, could decrease in an enhanced greenhouse gas environment.

During summer, the polar regions receive an almost constant influx of solar radiation. The low angle of the sun, however, causes a slow thaw of ice and permafrost. Areas of the high latitudes that are thawing during summer experience averaged surface temperatures near 0°C. This prevents significant increases in near-surface air temperatures during summer, even if the globally averaged climate warms considerably. This is consistent with the findings of Maxwell (1992), who showed little change in Arctic summer temperatures from 1961–1990 when other seasons had marked temperature changes.

Global winds

Differential heating of the Earth's surface, caused by latitudinal differences in incoming solar radiation, is the engine that drives global circulation. Figure 2.3 is a simplified schematic of this process, showing how large-scale flow might look in the Northern Hemisphere during winter if zonally averaged. Only troposphere winds are shown.

Direct sunshine at the equator warms the Earth's surface and causes the air to rise to the top of the troposphere. From there, it migrates toward the cooler polar regions. Eddies along the way (like the generalized Hadley cells shown in Figure

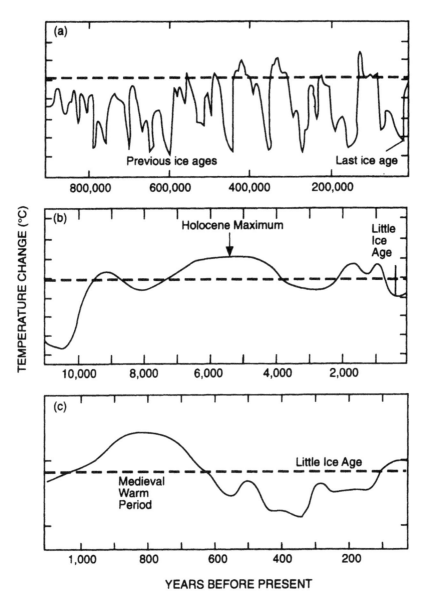

Figure 2.2. Schematic diagrams of global temperature variations since the Pleistocene on three timescales: (a) the last million years, (b) the last ten thousand years, and (c) the last thousand years. The dotted line nominally represents conditions near the beginning of the twentieth century. Adapted from IPCC (1990), figure 7.1. Reprinted with permission of Cambridge University Press and J. T. Houghton.

2.3) deflect the air somewhat. Strong lifting and convection near the equator cause abundant precipitation and a warm, moist climate. Hot, dry climates and desert environments develop where the air is deflected downward. Another area of lifting and convergence is where cool surface air from the poles meets warm air rising from the mid-latitudes. Here, a frontal zone is formed. Numerous storms generate around this front and are a principal source of moisture for the polar regions. Quasi-stationary features, like centers of low-surface pressure that persist near the Aleutian Islands, also entrain and transport moisture to the poles.

   A warmer atmosphere can retain and transport more moisture away from source regions before it condenses and precipitates. Therefore, more moisture in the system should be able to reach the polar regions during periods of global warming. Because of reduced solar input at the poles, water vapor that reaches the high latitudes cools and condenses. As it condenses, it gives off latent heat. This could enhance the effects of global warming in the high latitudes.

   It is difficult to determine the effect of enhanced greenhouse gases on storm patterns. A decrease in near-surface temperature gradients between the equator

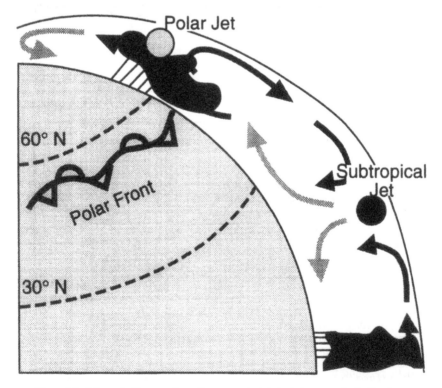

Figure 2.3. Schematic diagram of zonally averaged, global wind patterns in the Northern Hemisphere troposphere during winter.

and pole in the Northern Hemisphere could reduce the storm activity needed to maintain thermal equilibrium in the atmosphere. However, an increase in atmospheric moisture could also cause larger storms, because there should be more latent heat available to drive the system (Held 1993). Larger storms could retain more energy after crossing coastal mountains and bring more wind and precipitation to inland regions. The polar regions are currently dominated by winter temperature inversions (in which temperature increases with altitude), which cause strong gravitational stability in the atmosphere. In a warmer climate, surface heating could disrupt this stability, causing storm patterns to shift poleward (Held 1993). This effect, however, has been simulated in models only near regions of flat continents and simplified continental boundaries (Manabe et al. 1991) and may not apply to North America where the coasts are bounded by islands and mountains.

Oceans

A rise in sea-surface temperatures is expected in response to higher near-surface air temperatures, although the impact on northern coastal environments is complicated by ocean currents. Ocean currents carry water masses with different physical properties from one location to another. For example, the Alaskan Current, which flows counterclockwise around the Gulf of Alaska, brings relatively warm water from Japan northward along the British Columbia coast and maintains higher sea-surface temperatures within the Gulf than at similar latitudes in the western Pacific.

Coastal currents, such as the Alaskan Current, are strongly affected by river discharge. Fresh water from melting snow and ice has lower salinity and is less dense than residing seawater. The diluted, lighter water stays on the surface longer and can be more easily driven by wind stress. This allows greater intrusion of ocean water into coastal estuaries. Because melting snow and ice dominate river discharge in many northern rivers, changes in melting patterns could play an important role in the condition of coastal waters.

In the Pacific Ocean, the greatest influence on regional climate is the El Niño Southern Oscillation (ENSO), in which warm water moves back and forth across the tropical Pacific. For example, western streamflow decreases during positive ENSO events (Cayan and Peterson 1989), while lowland snowfall increases during negative ENSO events (Ferber et al. 1993). Because of the strong correlations between ENSO and regional climate, it is necessary to filter out its effects when looking for long-term trends; this may be possible by using CGCM simulations.

Sea level may rise from 20–60 cm during the next century as a result of global warming (IPCC 1990). Rising land masses, from tectonic influences or isostatic rebound from melting glaciers and permafrost, may counterbalance rising sea levels. However, areas experiencing land subsidence could suffer greatly from rising sea levels. Without compensating land rises, coastal regions along major river deltas like the Mackenzie could be affected greatly by an appreciable rise in sea level.

Albedo

An intriguing component of high-latitude warming is attributed to feedback mechanisms from changes in sea-ice and snow-cover patterns. Snow and ice strongly influence whether incoming solar radiation is reflected or absorbed by the Earth's surface. Seventy-five to 90% of incoming solar radiation is reflected back to the atmosphere by snow and ice. Only 25 to 50% is reflected by soil and vegetation (Berry 1981). The rest is absorbed and re-radiated at longer wavelengths, then trapped by the near-surface atmosphere. This feedback allows for small decreases in snow cover to significantly increase absorption of heat by land, which adds to the original heating.

If large areas of snow cover and sea ice disappear completely, there could be a dramatic change in snow albedo (percentage of incoming solar radiation that is reflected) and surface temperature. Warmer air temperatures could result in sea-ice formation and snow accumulation later in the autumn. In general, sea-ice albedo feedback is five to eight times stronger than the land snow-cover albedo feedback for a 2% increase in incoming solar radiation (Harvey 1988). Both effects are important in Arctic climate.

Lack of high albedo in the ocean due to the absence of sea ice during autumn would allow more solar heat to be absorbed and stored at the surface. Such an open water formation can result in 6° to 8°C of warming near the pack-ice edge (Lynch et al. 1995), thus causing feedback into the atmosphere and compounding the effects of warming.

Warmer air temperatures on land could generally reduce the areal extent of snow cover and cause it to accumulate later in autumn and melt earlier in spring (Groisman et al. 1994). Earlier snowmelt might also result from increased deposition of soot and particulates from "Arctic haze" onto the snow surface. Longer snow-free periods during autumn and spring could allow more solar heat to be absorbed by the exposed ground surface. This could substantially increase the growing season for vegetation.

Regional climatic patterns in northern North America

Topographic features will control regional climatic patterns despite any changes in globally averaged climate. For instance, much of the moisture from the Pacific Ocean will be blocked from reaching the North American interior by coastal mountain ranges in southern Alaska and British Columbia. In addition, Arctic Ocean air will often be prevented from reaching interior Alaska because the Brooks Range and Baffin Island mountains block warmer Atlantic air.

Based on topographic barriers, permafrost patterns, snow-cover patterns, and prevailing storm tracks, the northern latitudes of North America may be divided into four climatic zones: (1) Arctic coast, (2) subarctic interior, (3) Pacific coast, and (4) alpine transition. Approximate boundaries of these zones are shown in Figure 2.4. The climate in each zone is controlled chiefly by solar radiation, topography, albedo, and prevailing wind patterns.

Arctic coast

The north slope of Alaska, Northwest Territories north of Port Radium and Eskimo Point, Quebec's Ungava Peninsula, and the Canadian Archipelago are influenced by the proximity of the Arctic Ocean. The approximate southern boundary of this climatic region is associated with the margins of snow and ice. In winter, the boundary roughly coincides with the mean wintertime position of a major frontal zone, where cold air from the Arctic Ocean meets modified marine air that has intruded from the Pacific. During the summer, the frontal zone is usually displaced northward offshore, near the pack-ice boundary. Then the climatic boundary coincides with the southern extent of tundra. Winter temperatures average from −20° to −30°C; summer temperatures average from 1° to 10°C. Cold, dry winters promote the development of permafrost, which can be several hundred meters deep. Rivers and lakes stay frozen, typically from September to May. Most precipitation is solid and usually accumulates during autumn and late spring with brief intrusions of moist air from the south. A thin, dry snowpack often has a hard, wind-driven surface from Arctic cyclones.

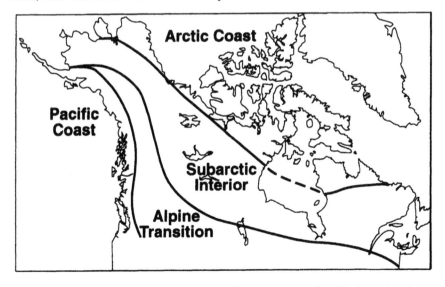

Figure 2.4. Approximate boundaries of four broad climatic zones in northern North America: Arctic coast, subarctic interior, Pacific coast, and alpine transition.

Along the Arctic coast and northern Canadian Archipelago, average winter temperatures could increase by 6° to 10°C, and average summer temperatures may increase from 0° to 2°C with doubled CO_2 (IPCC 1990; Maxwell 1992; Weller 1993). In the Arctic coastal region, the spring and autumn seasons are expected to lengthen during an enhanced greenhouse environment. This may be due to earlier melting of snow and ice during spring and later ice formation and snow accumulation during autumn. Little change in average summer temperatures is expected, because melting ice and permafrost modify daytime heating.

Higher humidities may accompany a warmer climate, because more moisture can persist in its vapor phase in a warmer atmosphere. However, temperatures should remain cool enough to help continue condensation and related precipitation. Therefore, slightly increased precipitation may occur in the Arctic coast region if climate becomes warmer (IPCC 1990, 1992; Maxwell 1992). Most precipitation should continue to accumulate during warmer autumns and springs.

Potential warming could also reduce the areal extent of sea ice. This could push the nutrient-rich sea-ice margin farther offshore, with potential effects on fish (Chapter 8) and marine mammal (Chapter 7) habitat. Sea ice could also be reduced to as little as 50% of its current thickness (IPCC 1990), allowing more heat flux from the ocean to pass through the ice. This would further intensify surface warming.

Temperature inversions are a unique aspect of the stable polar air mass in the Arctic. The inversion during winter is enhanced by substantial cooling over snow and ice surfaces. A summer inversion is associated with cooling over surface water deposits. A systematic reduction in midwinter surface inversions over the Arctic has been observed during the past few decades (Bradley et al. 1993). Occasional Arctic cyclones help to mix the atmosphere and temporarily clear out the haze. If Arctic storm frequency increases in a warmer climate, the frequency of inversions would continue to decrease, thereby enhancing near-surface warming.

Permafrost in the Arctic coastal region was formed thousands of years ago. It varies from a few meters to several hundred meters deep. Only the top meter or so of the active permafrost layer melts during summer; deeper layers remain frozen throughout the year. Continuous permafrost can exist in regions with average annual temperatures as high as −7°C. Because of the slow response of deeply buried ground ice to changes in air temperature, future warming should cause little degradation of Arctic permafrost. However, the active layer may refreeze more slowly during autumn and thaw more quickly during spring. Small changes in the active layer could also impact tundra vegetation and increase areas of related lakes and marshes (Chapter 5).

The southern boundary of the Arctic region in Alaska is defined by the Brooks Range. However, the southern boundary of the Arctic region in Canada has little terrain definition and could shift northward in a warmer climate. The current stability of permafrost in this zone should help maintain many of its distinct air mass properties during summer. However, if the pack-ice boundary were to remain offshore during winter, Arctic air mass characteristics could change sufficiently to shift the southern boundary slightly northward.

Subarctic interior

Interior Alaska, most of the Yukon Territory, the southern and central Mackenzie Basin, and northern parts of several provinces may be generally characterized as subarctic interior. The approximate southern boundary of the subarctic interior roughly coincides with the mean winter position of a major frontal zone. At this frontal zone, the cold continental air meets the relatively moist air masses that infiltrate from surrounding oceans. A continental influence causes huge extremes in seasonal temperatures. Winter temperatures average $-30°$ to $-40°C$, while summer temperatures average $10°$ to $20°C$. Most precipitation accumulates during summer showers; light winter snowfall occurs most frequently when Pacific storms penetrate inland.

A thin snow cover helps to insulate the ground and prevent large areas of permafrost from developing. This causes a discontinuous pattern, which is not as durable as the continuous permafrost of the Arctic coast. If greater magnitude Pacific storms occur in a warmer climate, they could push inland over the subarctic interior and cause an increase in winter precipitation and snow cover. Increases in snow depth could insulate the ground further and inhibit continuous winter freezing needed for permafrost formation. Increases in average temperature could limit ground freezing year round, causing a general disruption of permafrost in the subarctic. The Mackenzie Delta may be particularly susceptible to permafrost loss (Hare 1988).

Temperature inversions are strong and persistent in this zone. Arctic storms, which are weakened after crossing the coast or going through coastal mountains, do not have sufficient winds to scour away the temperature inversion. In addition to trapping pollutants, these inversions create extensive regions of winter fog. With an occasional influx of Pacific moisture, fog formations of ice crystals or "diamond dust" add to the eerie wintry scene.

Winter temperature could rise $8°$ to $12°C$ in Canada's northern interior during the next century (IPCC 1990, 1992; Maxwell 1992; Weller 1993). A more moderate rise in temperature ($4°$ to $8°C$) is predicted for the subarctic interior of Alaska. Unfortunately, GCMs do not have sufficient resolution to account for the effects of terrain barriers in central Alaska or intensified Pacific storms. Both help control winter temperature in the Alaskan subarctic, so GCMs probably underestimate temperature changes in central Alaska.

Although the subarctic interior is bounded by the Brooks and Alaska Ranges in Alaska, there are no similar terrain barriers in Canada. Each degree change in temperature can displace isotherms, and corresponding regional climate boundaries, a distance of 100 to 125 kilometers (IPCC 1990). This makes the southern boundary of Canada's subarctic interior one of the most unstable in the world.

Pacific coast

Climate along the Pacific coast is relatively mild. Cold air streaming through the Bering Sea is moderated by warm ocean currents in the Gulf of Alaska before reaching the coasts of southern Alaska and western British Columbia. Because the coasts are bordered by mountain ranges, moisture from the Pacific is lifted dramatically just after moving inland, producing abundant precipitation in coastal regions.

This region has undergone long-term variations in temperature (e.g., Morely and Dworetzky 1991), and historical data show short-term variations in precipitation (e.g., Cayan and Redmond 1994) in response to global circulation changes. There are preliminary indications that the Pacific coast region is less prone to drought than interior regions (Ferguson et al. in review), and may experience minimal change in precipitation with a warmer climate because coastal mountains trap precipitation efficiently. If Pacific storms are less frequent and storm tracks establish farther north, winter precipitation in southwestern British Columbia could be reduced while increasing precipitation could occur in southern Alaska. Increased magnitude storms, however, could cause a general increase in winter precipitation. Summer storm tracks could align farther north and push storms away from most of the region, causing a general decrease in summer precipitation.

Winter temperatures could increase by $2°$ to $5°C$ in this region during an enhanced greenhouse environment (IPCC 1990, 1992; Weller 1993). A $2°C$ increase in winter-averaged temperature could raise the seasonal snowline in this region by about 500 meters, affecting water supply, fish habitat, and riparian ecosystems. A general decrease in the areal coverage of mountain glaciers is also possible with global warming. Decreases should be most pronounced in areas where snow accumulation zones are significantly depleted. Tidewater glaciers, which strongly affect coastal marine habitat, generally respond more to the geometry of their fjord than to changing climate. Therefore, detecting a climate change signal in tidewater glaciers may not be possible during the next century.

Alpine transition

Few places on Earth have such a pronounced physical border as the Coast and Rocky Mountain ranges of North America. These mountains, which cover southern and southeastern Alaska, southwestern Yukon Territory, much of British Columbia, and southwestern Alberta, provide a boundary between warm marine air to the west and cold Arctic air to the east. These air masses move back and forth through mountain passes and over ridges in response to changes in surface pressure gradients and upper-level wind flow. Such conditions cause occasional disruption of the dominant climatic influences. For example, Prince George, British Columbia, has a modified marine air mass in winter with average minimum daily temperatures near $-7°C$, although minimum temperatures near $-28°C$ occur from Arctic incursions (Leovy and Sarachik 1991).

It is uncertain how the marine/Arctic air mass boundary will respond to climate change. If Arctic and subarctic temperature inversions become less frequent, the associated high atmospheric pressure should diminish. If there are less frequent Pacific storms in the next century, there could be fewer oceanic low centers. The loss of significant pressure gradients from the continental highs to the oceanic lows could reduce the surface flow that draws Arctic air from the subarctic interior into the alpine transition region. This could keep the region under moderate maritime conditions during winter and enhance effective warming.

Low-elevation snow cover and glaciers in the alpine transition region could be depleted in a warmer climate. Because the alpine transition region is away from moderating ocean influences, positive feedback from the loss of low-elevation snow should be relatively high. The enhanced winter warming could cause seasonal snow levels to rise 500 to 1500 meters in the alpine transition region during the next century. This loss of snow cover could devastate water resources in the region. For example, a 500-meter rise in seasonal snow line in the Columbia River Basin (southern part of the alpine transition region) could cause a 44% reduction in snow cover (Fleagle 1991).

Summary

There is considerable uncertainty concerning how a warmer climate would affect regional climatic patterns of northern North America. One of the major impacts of climatic warming, however, could be a shorter, less intensely cold winter season. As discussed above, the possible impacts include: (1) a decrease in daily temperature range; (2) less frequent Pacific storms with greater magnitude farther north; (3) more frequent Arctic storms; (4) less frequent inversions; (5) smaller mountain glaciers; and (6) altered lake, river, and sea-ice conditions, including early spring breakup, summer reduction, later autumn formation, and reduced winter thickness.

Possible changes in regional climatic patterns are summarized in Table 2.1. It is clear that the subarctic interior could experience the most dramatic changes, especially during late autumn and early spring when winter warming would be most concentrated. This region also has the most vulnerable boundary in Canada where few terrain barriers exist. The most significant changes in other regions are primarily due to expected depletions of snow and ice cover.

The level of potential climate change is unprecedented. By the end of the next century, northern latitude ecosystems could experience the warmest climate of this millennium. Although there is a high degree of uncertainty in the magnitude and rate of potential change, most scientists agree that a warming trend is likely with substantial increases in greenhouse gases. Potential climate change

Table 2.1. Regional changes expected in the climate of northern North America.

Region	Winter air temperature	Summer air temperature	Winter precipitation	Summer precipitation	Snow cover	Snow depth	Permafrost
Arctic coast	+6–10°C	+0–2°C	Slight increase	Slight increase	Little change	Slight increase	Longer active season
Subarctic interior	+8–12°C	+0–4°C	Slight increase	Increase showers Decrease steady rain	S. boundary shift up to 10° latitude	Decrease in south	Disruption
Pacific coast	+2–4°C	+2–6°C	Moderate increase	Slight decrease	Elevation boundary shift 500–1,000 m	Up to 20% increase	—
Alpine transition	+2–6°C	+2–6°C	Moderate increase	Slight decrease	Elevation boundary shift 500–1,500 m	Up to 20% increase	Decrease

will not be steady, and there will continue to be colder than "normal" seasons, although perhaps not as many. If there is a significant rise in globally averaged temperature in the next century, most of the warming is expected over high-latitude land areas.

References

Berry, M. O. 1981. Snow and climate. In *Handbook of snow*, edited by D. M. Gray and D. H. Male, 32–58. New York: Pergamon Press.

Bradley, R. S., F. T. Keimig, and H. F. Diaz. 1993. Recent changes in the North American Arctic boundary layer in winter. *Journal of Geophysical Research* 98(D5):8851–8858.

Cayan, D. R., and D. H. Peterson. 1989. The influence of North Pacific atmospheric circulation on streamflow in the west. In *Aspects of climate variability in the Pacific and the western Americas*, edited by D. H. Peterson, 375–397. Geophysical Monograph 55. Washington, DC: American Geophysical Union.

Cayan, D. R., and K. T. Redmond. 1994. ENSO influences on atmospheric circulation and precipitation in the western United States. In *Proceedings of the tenth annual Pacific climate workshop*, edited by K. T. Redmond and V. L. Tharp, 5–26. Interagency Ecological Studies Program Technical Report 36. Sacramento: California Department of Water Resources.

Ferber, G. K., C. F. Mass, G. M. Lackmann, and M. W. Patnoe. 1993. Snowstorms over the lowlands of western Washington. *Weather and Forecasting* 8:481–504.

Ferguson, S. A., M. Peterson, D. Lettenmaier, P. Hayes, and D. Jewett. Disturbance climatology of the Columbia River Basin. Columbia River Basin Assessment and Eastside Ecosystem Management Project. Seattle: USDA Forest Service. In review.

Fleagle, R. G. 1991. Policy implications of global warming for the Northwest. *Northwest Environmental Journal* 7:329–343.

Giorgi, F., and L. O. Mearns. 1991. Approaches to simulation of regional climate change: A review. *Review of Geophysics* 29:191–216.

Groisman, P. Y., T. R. Karl, R. W. Knight, and G. L. Stenchikov. 1994. Changes of snow cover, temperature, and radiative heat balance over the Northern Hemisphere. *Journal of Climate* 7:1633–1656.

Hare, F. K. 1988. Climate in the north: Prospects for change. In *Knowing the North: Reflections on traditions, technology and science*, edited by W. C. Wonders, 9–14. Occasional Publication 21. Edmonton: Boreal Institute for Northern Studies.

Harvey, L. D. D. 1988. On the role of high latitude ice, snow, and vegetation feedbacks in the climatic response to external forcing changes. *Climatic Change* 10:191–224.

Held, I. M. 1993. Large-scale dynamics and global warming. *Bulletin of the American Meteorological Society* 74:228–241.

Intergovernmental Panel on Climate Change (IPCC). 1990. *Climate change: The IPCC assessment*, edited by J. T. Houghton, G. J. Jenkins, and J. J. Ephraums. Cambridge: Cambridge University Press.

Intergovernmental Panel on Climate Change (IPCC). 1992. *Climate change 1992: The supplemental report to the IPCC scientific assessment*, edited by J. T. Houghton, B. A. Callander, and S. K. Varney. Cambridge: Cambridge University Press.

Karl, T. R., P. D. Jones, R. W. Knight, G. Kukla, N. Plummer, V. Razuvayev, K. P. Gallo, J. Linseay, R. J. Charlson, and T. C. Peterson. 1993. A new perspective on recent global warming: Asymmetric trends of daily maximum and minimum temperatures. *Bulletin of the American Meteorological Society* 74:1007–1023.

Leovy, C., and E. S. Sarachik. 1991. Predicting climate change for the Pacific Northwest. *The Northwest Environmental Journal* 7:169–201.

Lynch, A. H., W. L. Chapman, J. E. Walsh, and G. Weller. 1995. Development of a regional climate model of the western Arctic. *Journal of Climate* 6.

Manabe, S., R. J. Stouffer, M. J. Spelman, and K. Bryan. 1991. Transient responses of a coupled ocean-atmosphere model to gradual changes of atmospheric CO_2. Part I: Annual mean response. *Journal of Climate* 4:785–818.

Maxwell, B. 1992. Arctic climate: Potential for change under global warming. In *Arctic ecosystems in a changing climate*, edited by F. S. Chapin, R. L. Jefferies, J. F. Reynolds, G. R. Shaver, J. Svoboda, and E. W. Chu, 11–34. New York: Academic Press.

McBeath, J. H., ed. 1984. *The potential effects of carbon dioxide-induced climatic changes in Alaska: Proceedings of a conference*. School of Agriculture and Land Resources Management Miscellaneous Publication 83–1. Fairbanks: University of Alaska.

Meier, M. F., D. G. Aubrey, C. R. Bentley, W. S. Broecker, J. E. Hansen, W. R. Peltier, and R. C. J. Sommerville. 1985. *Glaciers, ice sheets, and sea level: Effect of a CO_2-induced climatic change*. Arlington, VA: National Technical Information Service, U.S. Department of Commerce.

Morely, J. J., and B. A. Dworetzky. 1991. Evolving Pliocene-Pleistocene climate: A North Pacific perspective. *Quaternary Science Reviews* 10: 225–37.

Schlesinger, M. E., and Z. C. Zhao. 1989. Seasonal climatic change introduced by doubled CO_2 as simulated by the OSU atmospheric GCM/mixed-layer ocean model. *Journal of Climate* 2:429–495.

Walsh, J. E., and R. G. Crane. 1992. A comparison of GCM simulations of Arctic climate. *Geophysical Research Letters* 19:29–32.

Weller, G. 1993. Global change and its implications for Alaska. In *Impacts of climate change on resource management in the North*, edited by B. Mitchell, 17–22. Department of Geography Occasional Paper 16. Waterloo, Ontario: University of Waterloo.

3

Demography and socioeconomics of northern North America: Current status and impacts of climate change

Jonathan P. Kochmer and Darryll R. Johnson

Predictions of future northern social and cultural conditions must be based on existing knowledge and components of contemporary social structure. Social trends in the distant future may be caused by circumstances that cannot be foreseen now. Considered individually, causal factors that can be identified may suggest contradictory futures and consequently the need for complex multivariate analyses. Such complex analyses (ignoring the problem of missing variables) would be possible only on the basis of what is known about the relationship of these variables today. Furthermore, small changes in complex systems of variables can greatly influence outcomes. Imagine how difficult it would have been in the 1920s to predict characteristics of modern North American culture and society. Therefore, this chapter does not attempt precise predictions or projections of the status quo into the next century. The subjects discussed are not inclusive of all possible factors relevant to future northern social conditions, but are presented as examples of potentially important variables that could drive social change.[1]

The national and international contexts of northern social change

Three determinants of demographics and economics of northern North America have been a challenging climate, resources that are patchily distributed or inac-

---

[1]Studies of future social change have the potential of being generative. Asserting that something might happen may lead to a "self-fulfilling prophecy" (e.g., a false rumor about a bank could cause depositors to withdraw funds and the bank to fail). In contrast, "self-defeating prophecies" are true statements that become false because of public response (e.g., "underdog effect," in which a projected loser in a political race wins because of sympathy votes). The future of the North is being shaped by prophecies concerning reserves of nonrenewable resources that might alleviate shortages and generate national revenues; many Canadians and Americans feel it is imperative to develop these resources. If accepted as inevitable, the negative conditions of Native peoples described in this chapter could exacerbate a tragic situation.

cessible to development, and a low carrying capacity for human populations. Changes in technology and pressures for resource development, combined with regional and global effects of a warmer climate, may drastically alter assumptions about northern economies and human populations. Simple extrapolation of current northern conditions assumes relative uninhabitability and cultural isolation, continued net export of raw materials and net import of processed materials, and economic development driven by extraction of raw resources, particularly fossil fuels. These assumptions need to be reconsidered in light of recent technological, sociopolitical, and ecological trends.

Technological trends

Globalization is increasing because of one-to-one communication or many-to-one telecommunications (e.g., mass media), and, more recently, many-to-many systems such as the Internet. Satellite-mediated systems will allow communication between any two locations on earth at a cost independent of physical location. About 20% of North Americans may be able to "telecommute" (work from computers remote from employers) at least part-time by the year 2000. Future workers could choose employers thousands of kilometers away, and many forms of employment currently implausible in the North may be possible. Another product of satellite-mediated telecommunications is distance education, already being delivered to communities in rural Alaska and northern Canada.

Electronic communication could play an important role in cultural change in remote rural areas, as children are exposed to the values, attitudes, and lifestyles of southern cities in the United States and Canada (LeBlanc 1994). Improved communication technologies may be important in articulating local political positions, as indigenous and rural people resist influence from the outside and form alliances with interest groups worldwide. Although outside influences may alter local economies, culture, and personal identities, communication technologies also will allow northern rural people to communicate with like-minded people throughout the world (e.g., Native Net on the Internet).

Current imperatives for development of the North derive from natural resources in the region, particularly petrochemicals. In the very long term, research on nonfossil fuel energy sources may yield economically viable solar or even safe fusion energy, reducing the need for fossil fuel and slowing petrochemical-driven development. Wind power is currently becoming established in Alberta (Tulley 1994), and could become an energy source of choice in other areas, (e.g., Keewatin in the Northwest Territories [NWT]). Tidal energy could become viable in areas with narrow fjords, such as southeastern Alaska or the Arctic Islands. Microwave transmission of electricity could reduce development costs in remote areas. In the intermediate future, however, pressures to develop northern petrochemical reserves will continue.

Although areas favorable to agricultural expansion because of global warming are limited in North America, biotechnological breakthroughs may allow farming of crops not currently grown in some soils and climates of the North. The greatest agricultural impact of warming may be conversion of marginal agricultural environments such as those in northern Alberta to more productive croplands.

Social and political trends

Ethnic groups throughout the world have been striving to create autonomous states with homogeneous populations. In the late twentieth century, this trend is most notable in Eastern Europe, Africa, and parts of Asia. Some systems theorists argue that the likelihood of chaotic behavior is directly proportional to the number of independent interacting units, and global political instabilities may be related to recently formed and diplomatically inexperienced nations.

Formerly agrarian developing nations are rapidly becoming industrial and post-industrial economies. This trend will continue in Asia and Central America and may begin in coastal African nations. Simultaneously, new regional arrangements such as the North American Free Trade Agreement and the European Union are encouraging greater competition in global markets. As long as Canadian and United States trade and budget deficits are high (caused by imports of manufactured goods and oil) and dollars devalued, there will be pressure to develop northern resources for export.

Canada may experience continued regional separatism in Quebec and possibly in British Columbia. In addition, Native groups are moving increasingly toward self-determination and self-governance. Over the last 30 years, Native peoples have been increasingly active in the courts and politics, especially relating to natural resource use (e.g., Notzke 1994). Establishment of Inuit self-rule in Greenland and the creation of Nunavut (a new Canadian territory under Inuit jurisdiction) in the NWT are inspiring other Native groups such as the Yukon Dene to seek greater political and governmental autonomy from the Canadian government. Separatist movements by indigenous peoples or other groups are unlikely to succeed in the United States within the next 50 years, although the movement toward Native tribal sovereignty in Alaska may have major implications for Native communities and ultimately for resource management in some areas. Decentralization and local control can be expected to increase in the intermediate future.

Environmental contaminants

Although a subsistence diet has significant dietary benefits (Chapter 10), global and regional pollution is rendering Arctic ecosystems potentially hazardous, because contaminants are concentrated in animals harvested by humans. Locally, Arctic ecosystems can have high concentrations of hydrocarbons, organochlorines, metals, and radionuclides contributed by effluents from military, mining, and drilling operations. Inuit women of northern Quebec have high organochlorine residue concentrations, in part from eating beluga (*Delphinapterus leucas*) fatty tissues (Dewailly et al. 1993). Finnish reindeer (*Rangifer tarandus platyrhyncus*) herders have very high levels of selenium and mercury (Luoma et al. 1992). Without stringent environmental regulation and monitoring, extractive development may increase such pollutants.

Globally, the polar atmosphere is a pollutant sink. Lichens may have concentrated radioisotopes from atmospheric fallout during the 1950s and 1960s, raising the possibility of even higher levels in ungulates such as caribou (*R. tarandus*) and musk-ox (*Ovibos moschatus*) (Thomas et al. 1992). Pollutants may increase as developing countries rapidly industrialize. These trends may not eliminate subsistence living, but they need to be considered when thinking about the future of traditional lifeways based on subsistence, given that the impacted people are few in number and have little political power.

Demographic information

This section presents past, current, and projected demographic data for Alaska, the Yukon Territory (hereafter the Yukon), and the NWT as case examples of population trends in the North. The basic elements of demographics are birth rates, migration, and mortality. Of these, migration is most likely to be substantially affected by global warming and causes the greatest uncertainty in long-term projections.

General population characteristics

Selected population characteristics of northern North America are summarized in Table 3.1 (U.S. Bureau of Census 1993; Statistics Canada 1993). As of 1991, approximately 635,000 people were counted by the United States and Canadian censuses in Alaska, the Yukon, and NWT. With an area of 5.2 million km², this is the least densely populated portion of North America (0.1 people/km²). Alaska has the highest density of these regions (0.4 people/km²); rural NWT is the least densely populated (0.01 people/km²).

Until recently, most northern residents lived in rural areas. Today, 67% of Alaskans and 65% of Yukon residents live in areas classified as urban (places with 2500 or more inhabitants). In Alaska, 52% live in Anchorage, Fairbanks, and Juneau. Twenty-six percent of NWT residents currently live in urban areas. However, increasing proportions of Yukon and NWT residents (including Natives) are residing in urban areas.

Recent censuses of the United States (1990) and Canada (1991) indicate that 20% of the inhabitants of Alaska, the Yukon, and NWT were counted as Natives. Regions of the North with greatest percentages of Natives are rural Alaska (28%)

Table 3.1. Total and Native population and population densities of the North, 1990–1991.

Region	Total population	Total (%)[a]	Region (%)[b]	Native population	Native (%)	Area (km²)	Number of people per km²
Alaska	550,043	86.6	100.0	86,125	15.7	1,518,776	0.36
Anchorage	226,238	35.6	41.1	14,569	6.4	—	—
Fairbanks	30,843	4.9	5.6	2,830	9.2	—	—
Juneau	26,751	4.2	4.8	3,462	12.9	—	0.12
Rural	178,808	28.1	32.5	49,385	27.6	—	0.05
Yukon	27,797	4.4	100.0	6,385	23.0	536,327	—
Whitehorse	17,925	2.8	64.5	3,110	17.4	—	0.02
All other	9,872	1.5	35.5	3,275	33.2	—	0.02
NWT	57,649	9.1	100.0	35,390	61.4	3,379,699	—
Yellowknife	11,753	1.8	20.4	2,965	25.2	—	—
Inuvik	3,206	0.5	5.6	1,695	52.9	—	0.01
All other	42,690	5.9	74.1	30,730	72.0	—	0.12
Total	635,489	100.0	—	127,900	20.1	5,434,801	

Sources: Data are from U.S. Bureau of Census (1993) and Statistics Canada (1993).

[a] Total (%) is the percentage of the entire region accounted for by that row.

[b] Region (%) is the percentage of that state or territory accounted for by that city or subregion.

and the entire NWT (61%). The NWT is the only major political region in Canada or the United States in which Natives form a majority.[2] The total population of these areas has grown rapidly, from around 67,000 in 1921 to 635,489 in 1990, an average annual rate of increase of 12%. Although Natives were probably under-counted in early censuses, their populations may have been smaller than today's by 70% (Lindsay 1980).

Table 3.2 summarizes fertility rates (children born per 1,000 ever-married women) in Alaska, the Yukon, and NWT. In Alaska, the Native fertility rate (3.14) is much higher than for Whites (1.90) or Asians (2.00). Highest fertility rates are among rural Natives in Alaska (3.43) and the NWT (4.43), almost double the gen-eral Canadian population. Urban Alaska Native rates (2.81) are lower than rural Alaska Native rates (3.40) but still much higher than Alaska Whites. Fertility rates in the Canadian population decrease as a function of maternal education. Thus, as a higher percentage of Natives attend college and inhabit urban areas, Native fertility rates could drop. Nonetheless, Native populations throughout the North, especially in rural Alaska, will probably have a greater natural increase than the general population.

Age pyramids of the Canadian population have a fairly even distribution of individuals in age classes up to 45. Alaska, the NWT, and to a lesser degree the Yukon, have large numbers of children and young adults (Figures 3.1 & 3.2) char-acteristic of populations experiencing more rapid growth. Given higher fertility rates and in-migration of young adults, it is not surprising that Alaska, the Yukon, and NWT have more individuals in younger age classes than United States and Canadian populations generally.

Table 3.2. Fertility expressed as children born per 1,000 ever-married women.

Region	Subgroup	All	Native
United States	All	2.10	2.62
Alaska	All	2.08	3.14
	Urban	1.91	2.81
	Rural	2.29	3.43
Canada	All	2.32	3.59
	No high school degree	3.56	N.A.
	College degree	1.53	N.A.
Yukon	All	2.13	3.00
NWT	All	3.08	4.43

Sources: Data are from U.S. Bureau of Census (1993) and Sta-tistics Canada (1993).

[2]There are diverse Native cultures in the North that evolved relatively independently for thou-sands of years. Anthropologists use three broad groupings to represent these people: Aleutians, who live primarily in coastal southwestern Alaska; Inupiats, who live on the coast and inland from north-western Alaska east to Greenland (called Inuit in Canada); and Athabascans, who live mainly inland in eastern Alaska, the central Yukon and west central NWT (see also Chapters 9 and 11).

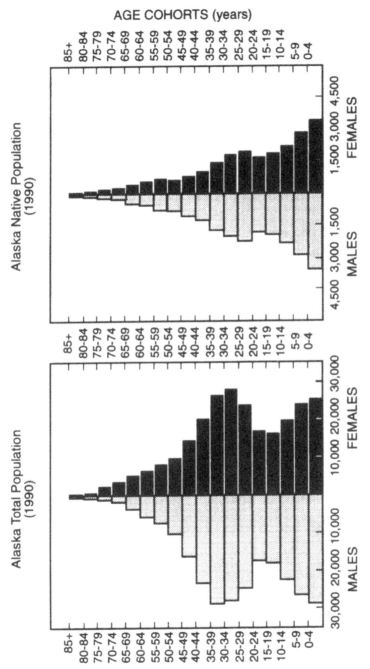

Figure 3.1. Age pyramids for Alaskan populations. Data are from United States Census (1990).

37

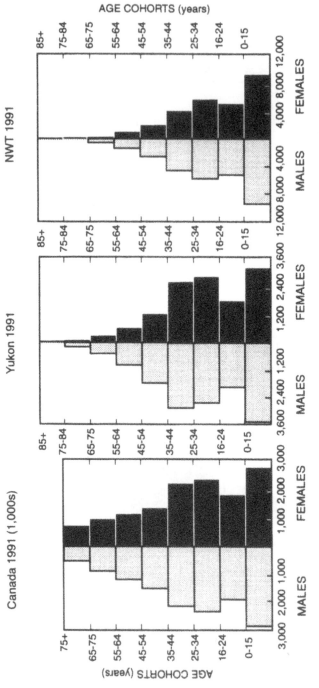

Figure 3.2. Age pyramids for Canadian, Yukon, and NWT populations. Data are from Statistics Canada (1991). Note that cohort categories are different than in Figure 3.1.

A characteristic of frontier societies is a high male-female sex ratio. Figure 3.3 shows historical sex ratios of the Yukon and NWT, compared with lagged historical sex ratios for Canada (lagged by 55 years) and British Columbia (lagged by 35 years). The real sex ratio trajectories of frontier areas overlap rather closely with the lagged historical sex ratios of the national data.

Migration

There is currently a net rural-to-urban migration of indigenous people in much of the North. Emigration is pronounced among indigenous women ages 20–35, leading to lower percentages of Native females in rural areas and high percentages of Native females in urban areas (Figure 3.4). An early survey in British Columbia suggests that the primary reasons for this movement are to seek education (36%) or employment (14%) (Stanbury 1975). One demographic implication of rural-urban migration may be a reduction in average Native fertility rates. In addition, many Native females may either intermarry into other groups or remain single.

Migration of rural residents into regional trade centers has implications for community planning and for subsistence management in protected areas in the

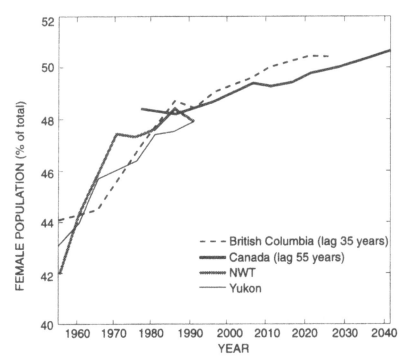

Figure 3.3. Historical sex ratios for Alaska, the Yukon, and NWT, compared with lagged historical sex ratios of British Columbia (lagged 35 years) and Canada (lagged 55 years). Data are from United States Census and Statistics Canada.

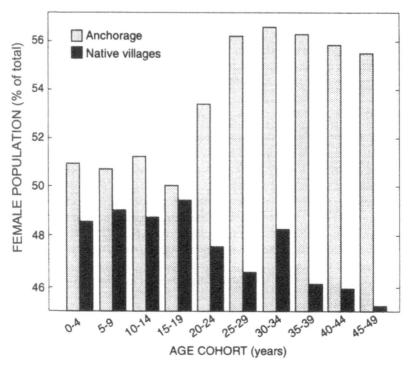

Figure 3.4. Sex ratios of indigenous populations in Alaska Native Village Statistical Areas and Alaskan urban areas, by age cohort. Data are from United States Census (1990).

intermediate future. Should these regional centers begin to attract migrants from outside these regions, there are additional implications for resource managers and local rural residents in the long term. At the community level, provision of housing and services is problematic because of declining state and federal subsidies. Local subsistence resources may be depleted, resulting in rural migrants going longer distances from cities with possible conflict between Native urban residents and Native rural residents. Federal protected area managers, who in Alaska must manage for local subsistence use priority, may face unprecedented challenges in balancing competition for scarce fish and game with other resource uses and in allocating resources to qualified residents (Chapter 17). Federally managed areas in Alaska will likely become more important to local rural people choosing to continue subsistence lifeways.

A smaller but significant pattern of movement in the North is periodic flow of non-indigenous people from cities into the country. Such movements were pronounced in the 1890s and early twentieth century gold-rush periods, and to some extent in Alaska during the 1960s "counterculture" period. Satellite-based telecommunications and educational opportunities in remote areas may renew this trend (particularly within driving or boating distance of urban areas). Certain areas with natural amenities, particularly in a warmer climate, could become havens for the semi-retired, the wealthy, writers, and artists.

Migration is the demographic variable most likely to change in response to global warming. Canada and the United States receive a huge number of immigrants annually, both in absolute numbers (230,781 to Canada and 1,827,167 to the United States in 1991), and as a percentage of national population (0.7% and 0.3% of national population size, respectively, from 1946 to 1986). Until the 1960s, most immigrants to the United States and Canada came from industrialized European nations. Asia now contributes the largest percentage of immigrants to Canada (53%), and Latin America the greatest percentage of immigrants to the United States (44%). In the last decade, most immigrants have been from nonindustrialized nations (Employment and Immigration Canada 1991; U.S. Department of Justice 1993).

Alaska is a state of recent immigrants. As of 1990, 66% of all Alaskan residents were born outside of the state. Even 10% of Natives had been born elsewhere. International immigration into Alaska has been increasing, both in numbers (250/year in the 1950s to more than 1,000/year by the late 1980s) and as a percentage of Alaskan population. Since 1980, international immigration to Alaska has been primarily from Asia (61%) followed by North America (18%) and Europe (13%).

The Yukon and NWT receive few international immigrants, with only 0.1% of Canadian immigrants from 1981 to 1991. If these regions received international immigrants at the current national rates, they would receive nearly 5,000 people per decade, significantly increasing long-term population projections. Given the small populations of these territories, large numbers of immigrants (probably into urban areas) could easily alter regional ethnic composition, with implications for local politics and culture and growing schisms between rural and urban areas (Employment and Immigration Canada 1991).

Historically, Canada and the United States have been destinations of choice for political refugees. With increased global political destabilization and the creation of new nation-states, this source of immigrants is likely to increase. Environmental degradation caused by deforestation, desertification, salinization of agricultural lands, and regional overpopulation is already producing a large number of "environmental refugees" (Suhrke 1994) that would create another large pool of potential immigrants. Global warming may exacerbate this situation.

Given recent trends in immigration to the United States and Canada, most new immigrants will be from less developed countries in Asia, Latin America, and possibly Africa. Depending on future sociopolitical climates, immigration policies might be restructured to encourage workers for target industries. Recent amendments to the Canadian Immigration Act will strengthen the ability of the federal and provincial governments to determine where immigrants settle. If the United States and Canada attempt to drastically limit immigration, the world community would probably object by claiming that northern North America is among the few viable destinations for refugees.

Mortality and morbidity

The high fertility rates of northern Natives are partly offset by high rates of infant mortality. For example, infant mortality rates of Canadian Natives are at least 2.5 times higher than in the general population (Lindsay 1980). Nonetheless, annual rates of Native population natural increase in Canada are still higher than the general population.

During initial periods of contact with European-Americans, introduced diseases killed a large proportion of Native populations throughout North America. In Canada, Natives have hospital admission rates 2.5 times higher than the general population. Although mortality rates from infectious diseases among northern indigenous peoples have fallen since initial contact, incidence rates of infectious diseases of Native Americans generally, and Alaskan Natives specifically, were higher than the general United States population as recently as 1970–1978 (Johnson and Exendine 1979).

Alaskans generally consume twice as much alcohol per capita as Americans from the lower 48 states (Smith and Hanham 1982). Levels of alcohol consumption and diseases and deaths caused by alcohol are even higher among indigenous peoples of Alaska and northern Canada (Andre 1979). Alcohol abuse is a contributing factor in four of the ten most frequent causes of Native American mortality in the United States—accidents, chronic liver disease and cirrhosis, homicide, and suicide. Recently, fetal alcohol syndrome has been increasing in Native communities, with estimated rates as high as 12% of the children in some northern communities (Streissguth 1994).

Among Alaska Natives, suicide rates have risen from a level equal to the United States population in 1960 to six times higher than the general United States population in 1985. Suicide rates of Alaska Native males aged 15–24 in 1985 were about 20 times higher than the general United States population. A statistically more robust interpretation can be made by measuring suicide rates over several years, and expressing rates in terms of standardized mortality rates (SMRs), or the ratio of a mortality factor observed in a cohort to the national average for that cohort. The highest suicide SMRs in Alaska (1978–1985) are for Native females 15–24 years of age (7 to 10 times national levels), followed by suicide SMRs for Native males 15–24 years of age (7 times national levels). Suicide SMRs for Alaska Native females 30–34 years of age drop to a level near United States average rates. In contrast, Alaska Native male suicide SMRs are more than two times the United States rates below 65 years of age (Forbes and Van Der Hyde 1988). The average annual suicide rate for Alaskan Native males from 1981–1990 was 274 per 100,000 compared to 44 per 100,000 for the same cohort of non-Natives (Berman and Leask 1994).

Living in the North can be dangerous, especially in the winter, and overall rates of fatal accidents in Alaska are double national rates. In the northern half of Alaska, the majority of accidental injuries leading to death are from firearms, while the most frequent source of injury reported is from falls (Johnson et al. 1992). Homicide rates for Alaska Natives are twice as high as Alaska Whites and are nearly equivalent to homicide rates for large cities in the lower 48 states. Most accidental deaths with firearms and many homicides and suicides are associated with alcohol consumption (Forbes and Van Der Hyde 1988).

Depletion of stratospheric ozone (probably due to chlorofluorocarbon accumulation) is correlated with increases in solar ultraviolet-B (UVB) radiation. UVB in the frequency range of 280–320 nm is known to induce basal and squamose skin cancers, as well as cause damage to the eyes (e.g., retinal degeneration and cataract production) (Bentham 1993). UVB radiation effects are especially pronounced toward the poles, and several studies have observed increased levels of nonmelanoma skin cancers in higher latitudes (Henrickson et al. 1990; Schaart et al. 1993). Increased UV radiation may also have generally detrimental effects on the immune system (Jeevan and Kripke 1993) and may activate human immunodeficiency viruses (Bentham 1993).

Population projections

Official population projections for the Far North do not consider potential effects of global warming. Table 3.3 summarizes recent population projections by the United States Census Bureau, Statistics Canada, and the United Nations. The total population of Canada is projected to reach 34,081,500 to 39,387,500 by the year 2016. Projected population sizes for the Yukon and NWT in 2016 range from 35,700 to 51,600 and 87,000 to 106,000, respectively.

If one projects world population from the 2020 figure of 8 billion (Table 3.3) at a 1.4% rate of annual increase to the year 2075, world population is estimated at about 17.2 billion. United States population projected from 2020 at a 0.5% annual increase is about 380.9 million; using a 1.5% rate of increase results in a projected population of 823.8 million.

For Alaska, using a 1% annual increase beginning in 2020 with the low estimate (Table 3.3), the projected population is about 1.5 million in 2075. If one assumes a 2.8% annual rate of increase in Alaska to 2075, the projected population is 4.6 million. Beginning in 2016 with the low estimates and projecting the Table 3.3 figures to 2075 at a 1% annual increase results in estimates of 156,488 for NWT and 64,214 for the Yukon. Using a 3.4% rate of increase for NWT and the Yukon, the population in 2075 is 762,110 and 370,989 respectively. These projections are offered not as predictions, but rather to illustrate that there may be substantial world and regional population growth during the next century; very different outcomes result from small changes in average annual growth rates.

Table 3.3. *Past, current, and projected populations.*

Region	1920s population[a]	Current population[b]	1920s–1990s yearly increase[c]	(%)	Projected population	Year (series)[d]	1990–2020 yearly increase	(%)
World		5.6 billion			8.0 billion	2020 (N.A.)	79.0 million	1.4
United States	106.5 million	249.9 million	2.0 million	1.9	289.6 million	2020 (Low)	1.3 million	0.5
					363.2 million	2020 (High)	3.8 million	1.5
Canada	8.8 million	27.3 million	264,412	3.0	34.1 million	2016 (Low)	271,386	1.0
					39.4 million	2016 (High)	483,625	1.8
Alaska	55,036	550,043	7,858	14.2	849,000	2020 (Low)	9,965	1.8
					1.0 million	2020 (High)	15,342	2.8
NWT	8,143	57,649	707	8.7	87,000	2016 (Low)	1,174	2.0
					106,000	2016 (High)	1,934	3.4
Yukon	4,157	27,797	338	8.1	35,700	2016 (Low)	320	1.2
					51,600	2016 (High)	952	3.4

Sources: Data are from U.S. Bureau of Census, Statistics Canada, and United Nations (1992).

Note: Rates from the 1920s to the 1990s should be viewed with some caution, because Natives were grossly underestimated in the 1920s censuses; Canadian Native populations in the early twentieth century were about one-third of current levels (e.g., Lindsay 1980). Official projected population growth rates for the North are far below observed historical growth rates.

[a]Past populations are for 1920 (world, U.S., and Canada) and 1921 (Canada, Yukon, and NWT).

[b]Current populations are for 1994 (world), 1990 (U.S. and Alaska), and 1991 (Canada, Yukon, and NWT).

[c]Yearly increases (%) are the percentage increase over the period, divided by the number of years in the period.

[d]Year (series) refers to the year to which the projection was made and the lowest and highest-projected population sizes.

Revised demographic projections

Official population projections are inconsistent with some important historical patterns. Although projected annual growth rates of Canada as a whole are 0.33 to 0.60 times rates observed from the 1920s to the 1990s, projected growth rates of the Yukon and the NWT are as little as 0.20 times inferred historical rates. Alaska's projected rate of increase is even more at odds with historical rates. The projected absolute increases appear far below what could be experienced through immigration in a "more habitable climate" scenario. Furthermore, there is evidence that Alaska Native populations are increasing above expectation, due in part to better health care and reductions in infant mortality. The impact is now being felt in housing shortages in selected remote areas (Richardson and Pickett 1994).

Current population densities of northern ecozones can be compared with population densities currently observed in southern portions of the same ecozone (Table 3.4). Admittedly, southern and northern parts of an ecozone can differ considerably in biotic and abiotic conditions. However, this approach is useful in considering how carrying capacities (the ability of the land and resources to support human populations) of these ecozones might change as a warmer climate makes northern regions more similar to their current southern counterparts.

For example, in 1991, 545,691 people inhabited the 786,489 km^2 of Boreal Plains of Canada apart from the Yukon and the NWT (0.69 people/km^2), while 34,344 km^2 of Boreal Plains in the Yukon and NWT supported 1,334 people (0.04 people/km^2) (Statistics Canada 1994). If population densities of the Canadian Boreal Plains were projected to the Yukon and NWT, these Boreal Plains might support 26,199 people, instead of the current 1,334. These "ecozone projections" suggest that many more people would live in the Boreal Plains and Southern and Northern Arctic ecozones of the Yukon and NWT. This projection is nearly identical to the highest population sizes projected by the Canadian Census for the Yukon and the NWT in 2016 (census projection: 157,600; ecozone projection: 161,558).

Global warming and northern economies

Economic sectors can be divided into three categories: extractive, based on agriculture and natural resources; industrial, using energy and machines to produce goods; and post-industrial, based on processing and distribution of information. Although roughly corresponding to a historical progression, these sectors are present in all modern societies in varying degrees.

Cash economies of the North have been traditionally based on extractive sectors such as mining, forestry, and harvesting of fish and fur-bearing animals. Industrial sectors have been difficult to sustain without substantial subsidies because of small populations, long distances from potential markets, and unfa-

Table 3.4. Observed population sizes and densities in Canada by ecozone.

Ecozone	Canada (except Yukon and NWT)			Current Yukon and NWT		Projected Yukon and NWT population
	Area (km²)	Population	Density	Area (km²)	Population	
Tundra Cordillera	0	0	—	282,346	264	264
Boreal Cordillera	120,607	2,177	0.018	259,506	26,199	26,199
Boreal Plains	786,489	545,691	0.694	34,344	1,334	23,835
Taiga Plains	107,244	1,014	0.009	476,964	16,994	16,994
Taiga Shield	914,695	37,905	0.041	470,308	18,238	19,283
Hudson Plains	388,961	10,857	0.028	3,121	0	87
Southern Arctic	191,417	7,238	0.038	737,058	9,731	28,008
Northern Arctic	67,105	2,216	0.033	1,359,619	10,665	44,867
Arctic Cordillera	17,659	0	—	242,597	2,021	2,021
Total	2,594,177	607,098	—	3,865,863	85,446	161,558

Sources: Data are from Statistics Canada (1994).

Notes: The last columns indicate the number of people that would inhabit these ecozones in Yukon and the NWT if densities were equal to densities observed in these ecozones in the rest of Canada. Populations for those ecozones of the Yukon and NWT in which the densities are already greater than in Canada as a whole are maintained at observed levels.

vorable climate. Consequently, the North has historically been a net importer of manufactured goods. Like lesser developed nations, the North might leapfrog industrial development and become a post-industrial service and information-based economy.

Northern economies have repeated "booms and busts" in response to cycles of resource discovery and depletion and supply and demand in world markets. This has occurred with furs and mining and will probably take place with petrochemicals and fisheries. A dual economy with wages, cash, and entitlements dominates in cities and towns, and mixed cash and subsistence economies prevail in rural areas.

Resource extraction

Petrochemicals have been a dominant contributor to northern economies since the early 1970s. Additional deposits have been discovered in Alaska, the Yukon, the NWT, the Arctic Islands, and Greenland. Global warming will generally reduce costs of prospecting, extraction, and delivery. Petrochemical activity in the North has been mainly extractive, but global warming may make refineries economically viable in southern Alaska or possibly even the Arctic Ocean (Delphi 1983).

The North has extensive deposits of zinc, lead, thorium, and uranium; large areas of diamond-bearing kimberlite have been discovered in the NWT in the past five years. The physical and social effects of mining can be long lasting. Towns, roads, water-polluting mine tailings, and cultural changes can long outlast mines that prompted their development. Mining activity has attracted indigenous peoples into towns and the wage economy; for example, the copper-nickel mines in Rankin Inlet on the west coast of Hudson Bay created one of the major settlements of the eastern Arctic by 1971. After the mine closed, the Inuit people stayed, largely supported by government payrolls and transfer payments, with a substantially reduced reliance on subsistence living (Foster 1972).

Forestry

Forestry activities will increase as main roads proliferate through the North (Figure 3.5). American, Canadian, and Japanese logging concerns are now harvesting forests from Alaska to northern Quebec. Changes in the distribution, abundance, and vigor of northern forests could greatly alter the long-term economic viability of commercial forestry. Given relatively slow rates of forest succession into newly suitable areas and slow tree growth at northern latitudes, many regenerated forest areas may not be available for subsequent harvest for over 100 years after logging.

Agriculture

Agriculture in the North is currently limited to relatively small farms producing hay, oats, cabbages, potatoes, and some livestock. In 1991, estimated farm

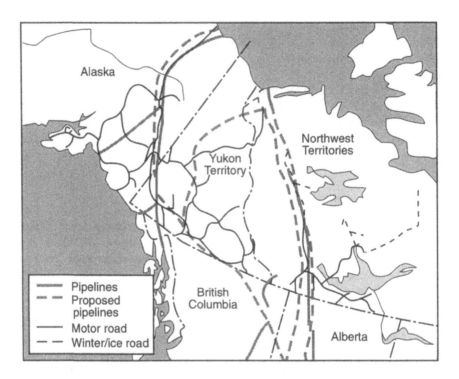

Figure 3.5. Existing and proposed roads and pipelines in northern North America.

receipts of Alaska amounted to $26 million (Alaska Agricultural Statistical Service 1993) and $2 million each in the Yukon and NWT (Statistics Canada 1992). Agricultural activities have been moving northward in British Columbia, Alberta, and Saskatchewan. A warmer climate could facilitate increased agricultural production (Smit 1989) in the southern Yukon, southern NWT, and current Alaskan agricultural areas such as the Matanuska Valley. Greenhouses now account for about 50% of the total agricultural sales of Alaskan agriculture (Alaska Agricultural Statistical Service 1993) and could become important in the Yukon and NWT as well. Nonetheless, the North is unlikely to become a net exporter of agricultural goods (Delphi 1983), and a warmer climate favors agricultural expansion more in Russia than in Alaska and Canada.

Fishing

Commercial fisheries of species such as salmon (*Oncorhynchus* spp.), halibut (*Hippoglossus stenolepis*), pollock (*Pollachius chalcogrammus*), and king crabs (*Paralithodes camtschatica*) are a major source of income in Alaska, generating almost 1.6 billion dollars in 1992. Fish are also an important source of food for subsistence harvesters (Chapter 10). The potential effects of a warmer climate on northern anadromous fish could be far-reaching for both commercial and subsistence harvesters (Chapter 8). A few freshwater fisheries may be significant contributors to locally growing markets, (e.g., in Yellowknife, NWT), but commercialization of freshwater fish is not widespread.

Fur harvesting

Although furs were the lifeblood of rural northern cash economies from the 1700s to the mid-twentieth century, the Western market for furs has been hurt by animal rights organizations and reduced demand for furs (Chapter 11). It is conceivable that this trend will reverse in the future and that demand for furs as luxury items will increase, especially with growth of non-Western markets such as China. Just as this chapter was going to press, an item from the Associated Press indicated that the demand for furs in China and Hong Kong is booming. However, trappers of northern wild fur will continue to face stiff competition from fur farms, and it is unlikely that wild furs will return as a major force in northern economies. A recovery in the fur market on a local basis, especially with value-added activities, may remain important to people living subsistence lifestyles.

Alternative energy development

Reduction of CO_2 emissions from fossil fuels is a potential incentive for development of alternative forms of energy. Hydroelectric concerns will argue that dams are nonpolluting and reliable sources of energy. Although dam construction provides temporary employment for large numbers of skilled and unskilled workers, hydroelectric development would likely lead to the destruction of areas supporting subsistence resources. Therefore, major hydroelectric projects are likely to face Native opposition, as in the case of James Bay in Quebec. If energy costs escalate in the face of pressures to reduce CO_2 emissions, the "correct" environmental position on these projects will be elusive.

Infrastructure

Extractive industries will create incentives for expansion of currently sparse transportation infrastructure throughout the North (Figure 3.5). New transportation corridors will mainly be roads, but railways to the Brooks Range or connecting Alaska to the lower 48 states are possible (Delphi 1983). Existing main highways such as the Alaska Highway and the Mackenzie Valley Road will probably be significantly upgraded (e.g., IBI Group 1990). An east-west corridor from Fairbanks to Hudson Bay, by way of Whitehorse or Norman Wells, is conceivable by the mid-twenty-first century. Isolated villages lying along transportation corridors would experience economic and population growth, and access roads for resource extraction could develop from main corridors. In regions with deep permafrost, warmer temperatures may reduce surface stability for overland transport.

A warmer climate would facilitate ship transport along much of the northern coast of North America for longer periods of time over longer distances. There may be longer ice-free periods in the Arctic Ocean, Hudson Bay, and near the Arctic Islands, as well as major rivers like the Mackenzie and Yukon. Throughout northern oceans there may be increased free-ranging pack ice, increased calving of icebergs from tidewater glaciers, and larger expanses of rough, open water.

Services and manufacturing

The largest and most rapidly growing employment sectors in Alaska and northern Canada are services and state and Federal governments, especially government-sponsored health care and social services. If employment in primary industries increases, there will be a need for services in secondary business such as retail and sales. The retail sector currently has the second highest level of employment in the Alaskan economy, and retail centers are expected to be a growth industry in Alaska (Safir 1991).

In the future, there will be significant growth in all aspects of telecommunications for business, education, and leisure. Notably, the Inuit Broadcasting Corporation has been working to establish a film and video network for indigenous peoples (Marks 1994) and intend to use this medium to improve conditions for Inuit and to promote a positive image for indigenous peoples generally (Weatherford 1992). In Alaska, Jeanie Green's Native-owned, produced, and managed television show, *Heartbeat Alaska*, is broadcast by several Canadian stations and 30 stations in the lower 48 states.

With the exception of southern Alaska, manufacturing has played a small part in northern economies, because small populations and large distances between population centers rendered local production more costly than imports. However, populations in some northern cities could increase enough to support manufacturing in some industries that use locally-harvested resources.

Tourism and marketing of indigenous culture

The Far North (Alaska in particular) possesses a mystique for Canadian and American Southerners as well as for Japanese and European travelers. Alaska has the fourth highest percentage of all states in tourism trade employment (Edmonson 1994). Approximately 861,000 visitors came to Alaska in 1994, a 10% increase over 1992. The tourism industry generates $1 billion a year for the state's economy.

Alaska appeals to tourists because of its image as a pristine wilderness and a frontier. It may increasingly be attractive because of its indigenous cultures, and the tourist industry may attempt to parlay the cultures of Native peoples into attractions (Elliott 1993). Some Native groups increasingly see tourism as a source of jobs and cash income. Several Alaska Native regional and village corporations are assessing the potential for tourism in local communities. [A word of caution may be in order because in many analogous situations ethnotourism has been destabilizing to local cultures (Price 1994)]. Big game hunts are currently being advertised in the United States by Native hunters and trappers associations in three communities through NWT Economic Development and Tourism (see Big Game Adventures 1995). At least one Dogrib outfitter in the NWT is offering highly acclaimed caribou hunts for nonlocal hunters out of Yellowknife. A warmer climate may increase the attractiveness of the North as a tourist destination, and investment parameters could change because of a potentially longer tourist season.

Summary

Northern North America has undergone rapid change during the twentieth century. With global warming (and imminent technological and sociopolitical trends), rates of change will accelerate during the twenty-first century. The North has been sparsely inhabited and home to cultures for whom culture was a form of technology. Like all places, the North of the future will probably be increasingly occupied by the culture of technology. The North has experienced rapid population growth throughout the current century and will probably continue such growth (especially from immigration) into the next century.

Average rates of natural increase of populations in the North are likely to decline with urbanization and other factors. These demographic changes suggest probable changes in state and local politics and in the collective definition of local self-interest in natural resource policy. In the future, there may be an even sharper contrast between rural and urban attitudes toward natural resource management policy.

It is difficult to imagine northern scenarios in which pressures for economically viable resource development will decline. A warmer climate will increase resource extraction activities, leading to growth of all other economic sectors and encouraging immigration. The magnitude and complexity of these changes will create challenging situations for rural peoples; natural resource managers; and local, state, provincial, and national governments.

In the future, northern people will face situations with world views we cannot anticipate, and perhaps with knowledge and technologies we cannot imagine. Given the probable growth in population alone, the management of subsistence resources in most protected areas will be challenging and increasingly contentious. The greatest challenge of "futures" studies is to transcend contemporary beliefs and to expect the unexpected.

References

Alaska Agricultural Statistics Service. 1993. *Alaska agricultural statistics.* Palmer: Division of Agriculture, Alaska Department of Natural Resources.

Andre, J. M. 1979. *The epidemiology of alcoholism among American Indians and Natives.* Albuquerque, NM: Indian Health Services.

Bentham, G. 1993. Depletion of the ozone layer: Consequences for non-infectious human diseases. *Parasitology* 106:39–46.

Berman, B., and L. Leask. 1994. Violent death in Alaska: Who is most likely to die? *Alaska Review of Social and Economic Conditions.* Anchorage: Institute of Social and Economic Research, University of Alaska.

Big Game Adventures. 1995. Cochrane (Alberta) Big Game Adventures Limited. *Big Game Adventures,* winter issue.

Delphi. 1983. *A Delphi forecast of Alaska's development: The year 2000 and beyond.* Report by Alaska Pacific University. Anchorage: Alaska Department of Commerce and Economic Development.

Dewailly, E., P. Ayotte, S. Bruneau, C. Lalibert, D. C. Muir, and R. J. Norstrom. 1993. Inuit exposure to organochlorines through the aquatic food chain in arctic Québec. *Environmental Health Perspectives* 101:618–620.

Edmonson, B. 1994. Tourist towns. *American Demographics* 16:60–61.

Elliott, C. 1993. Last frontier offers insights into culture, history of region. *Travel Weekly* 52:42.

Employment and Immigration Canada. 1991. *Immigration Statistics.* Hull, Quebec: Electronic Information Management, Citizenship and Immigration.

Forbes, N., and V. Van Der Hyde. 1988. Suicide in Alaska from 1978 to 1985: Updated data from state files. *American Indian and Alaska Native Mental Health Research* March: 36–55.

Foster, T. W. 1972. Rankin inlet: A lesson in survival. *The Musk-Ox* 10.

Henriksen, T., A. Dahlback, S. H. Larsen, and J. Moan. 1990. Ultraviolet-radiation and skin cancer. Effect of an ozone layer depletion. *Photochemistry and Photobiology* 51:579–582.

IBI Group. 1990. The implications of long-term climatic changes on transportation in Canada. *Climate Change Digest* CCD 90–02. Ottawa: Atmospheric Environment Service, Environment Canada.

Jeevan, A., and M. L. Kripke. 1993. Ozone depletion and the immune system. *Lancet* 342:1159–1160.

Jenish, D. 1994. Destination Canada: Even before the election, thousands of white South Africans had voted with their feet. *Macleans* 107:28–29.

Johnson, E. A., and J. N. Exendine. 1979. *Illness among Indians and Alaska Natives.* Washington, DC: U.S. Department of Health, Education, and Welfare.

Johnson, M. S., M. A. Moore, and R. D. Kennedy. 1992. Injuries in the Alaskan Arctic. *Arctic Medical Research* 51:45–55.

LeBlanc, L. 1994. Canada's aboriginal musicians seek mainstream recognition. *Billboard* 106:1–2.

Lindsay, C. 1980. The Indians and Metis of Canada, 173–182. *Perspectives III.* Ottawa: Statistics Canada.

Luoma, P. V., C. Nayha, L. Pyy, H. Korpela, and J. Hassi. 1992. Blood mercury and serum selenium concentrations in reindeer herders in the arctic area of northern Finland. *Archives of Toxicology Supplement* 15:172–175.

Marks, L. U. 1994. Reconfigured nationhood: A partisan history of the Inuit Broadcasting Corporation. *Afterimage* 21:4–8.

Notzke, C. 1994. *Aboriginal peoples and natural resources in Canada.* North York, Ontario: Captus Press.

Price, M. 1994. A fragile balance: The impact of adventure tourists on the ecological balance of remote areas. *Geographical Magazine* 66:33–34.

Richardson, J., and A. Pickett. 1994. Rural housing reports homes lag seriously behind current, future needs. *Tundra Times,* December 1.

Safir, A. 1991. Mauling job growth estimates. *Alaska Business Monthly* 7:12.

Schaart, F. M., C. Garbe, and C. E. Orfanos. 1993. Disappearance of the ozone layer and skin cancer: attempt at risk assessment. *Hautarzt* 44:63–68.

Smit, B. 1989. Climate warming and Canada's comparative position in agriculture. *Climate Change Digest* CCD 89–01. Ottawa: Atmospheric Environment Service, Environment Canada.

Smith, C. J., and R. Q. Hanham. 1982. *Alcohol abuse: Geographical perspectives.* Washington, DC: Association of American Geographers.

Stanbury, W. T. 1975. *Success and failure: Indians in urban society.* Vancouver: University of British Columbia Press.

Statistics Canada. 1992. *Agricultural profile of Canada, part I.* Catalogue 93-350. Ottawa: Ministry of Industry, Science, and Technology.

Statistics Canada. 1993. Age, sex, marital status, and common-law status. 1991 Census Technical Reports, Reference Products Series, Catalogue 92-352E. Ottawa: Ministry of Industry, Science, and Technology.

Statistics Canada. 1994. *Human activity and the environment.* Catalogue 11-509E. Ottawa: Ministry of Industry, Science, and Technology.

Streissguth, A. P. 1994. Fetal alcohol syndrome: Understanding the problem; understanding the solution; what Indian communities can do. *American Indian Culture and Research Journal* 18:45–83.

Suhrke, A. 1994. Environmental degradation and population flows. *Journal of International Affairs* 47:473–496.

Thomas, D. J., B. Tracey, H. Marshall, and R. J. Norstrom. 1992. Arctic terrestrial ecosystem contamination. *Science of the Total Environment* 122:135–164.

Tulley, A. 1994. The age of wind power dawns in Alberta. *Canadian Geographic* 114:12.

United Nations. 1992. *Long-range world population projections: Two centuries of population growth, 1950–2150.*

U.S. Bureau of the Census. 1993. *1990 census of population, social, and economic characteristics.* Washington, DC: Government Printing Office.

U.S. Department of Justice, Immigration and Naturalization Service. 1993. *1992 statistical yearbook of the immigration and naturalization service.* Washington, DC: Government Printing Office.

Weatherford, E. 1992. Starting fire with gunpowder. *Film Comment* 28:64–67.

PART II

Predicting Environmental Change

Northern landscapes require an adjustment in spatial scale and world views by southern visitors. The immensity of northern ecosystems gives them the illusion of stability and resilience in the face of climate change or human impacts. Biological data, cultural information, and the knowledge of elders all tell us that nothing could be further from the truth.

This section emphasizes that a warmer climate could alter biological systems at a very large scale, shifting entire biomes northward. A critical feature of boreal forests and tundra is their capacity to retain carbon. These two ecosystem types contain over a quarter of the Earth's terrestrial carbon in their cold soils. But warmer temperatures may result in a rapid breakdown of organic material and a net release of carbon to the atmosphere, providing a strong positive feedback to greenhouse warming. This fact alone illustrates the importance of northern systems at the global scale.

Nowhere on Earth are animals more important to human existence than in the North. Large mammals, fish, and other animals provide food for the stomach and sustenance for the spirit. Changes in the distribution and abundance of animal populations affect food supplies, local economies, and the cultural fabric of rural communities. The common theme of the discussions presented here is that even short-term fluctuations in weather can have a dramatic impact on large terrestrial mammals, marine mammals, and fish. Superimposing long-term climatic trends on smaller-scale climatic variation could spell disaster for some species. Or it could benefit some species. Or more likely it would shift the distribution of species. Observations of the response of animal populations to short-term climatic phenomena can provide only general guidance for the future.

Will animal species be able to track shifting habitats and maintain viable populations in the future? Will there always be caribou to hunt and salmon to harvest? Will tree line move northward and displace the tundra? It depends on the magnitude and rate of warming. We know that fluctuations in the Earth's climate have effectively rearranged the distribution of plant and animal species since the last Ice Age. It is reasonable to assume that a warmer climate would rearrange ecosystems again. As in the past, some species will flourish and others will decline. It remains to be seen whether *Homo sapiens* will be a winner or loser.

4

Modeling potential impacts of climate change on northern landscapes

Patrick N. Halpin

Estimating the potential impacts of climate change on ecological function, biodiversity, and human subsistence use for northern North America encompasses a wide range of complex systems over long time horizons. Climatic changes may alter resource availability to residents of the region through direct physiological stress on key species, alteration of critical habitats, change in seasonality, or extreme climatic events. In addition, shifts in human resource use related to climatic change may cause secondary impacts on resource use and protection. Logical methods for assessing the sensitivity of northern systems to climate change are needed to understand the risks to natural resources and to initiate effective management responses.

The models used to study sensitivity of climate to changing levels of greenhouse gases (atmospheric gases that are transparent to short-wave radiation and absorb long-wave radiation; examples include water vapor, carbon dioxide [CO_2], and methane) were developed not to forecast spatial predictions of future conditions, but to be used as theoretical experiments of geophysical processes (Chapter 2). These general circulation models (GCMs) use coarse spatial grids to define the fluid dynamics of global climate (Manabe and Stouffer 1986; Schlesinger and Zhao 1988). Ecological processes and subsequent land management decisions are generally considered at much finer spatial scales. The complex interactions of changing surface conditions and biogeochemical fluxes may exert significant controls on atmospheric processes not fully represented in current GCMs (Henderson-Sellars 1990).

Little is known about the physiological and ecological limits to species distributions and biological processes of northern landscapes. The rate of change and path of ecological responses to changing climatic conditions (Chapin et al. 1992) must be considered when exploring climatic and ecological scenarios. These climatic and ecological change scenarios are best viewed as heuristic devices (tools

I thank Nate Stephenson for helpful comments and suggestions on this chapter.

that help to learn or understand) that can be used to establish approximate boundaries of potential change.

In this chapter, climate change scenarios representing doubled carbon dioxide atmospheric conditions are mapped for North America and assessed in terms of their potential impacts on natural ecosystems and protected areas. General implications of these continental scale changes are then suggested for biotic resources of northern regions.

Two methods of analysis emphasize important differences in the interpretation of potential climate change with respect to short- and long-term ecological implications. Changes in climatic zones correlated with current vegetation distributions are used to identify potential long-term environmental stresses on continental-scale biogeographical patterns (Emanuel et al. 1985; Leemans and Halpin 1992; Prentice et al. 1992; Solomon 1992). This method identifies areas of potential ecosystem stress by tracking the change in climate as it relates to ecosystem patterns.

In contrast, an index of effective moisture change is used to identify immediate changes in the direction and magnitude of important climatic conditions predicted by the GCMs (Prentice et al. 1992; Halpin 1993). This method does not correlate specific climates to vegetation types, but maps important characteristics of the climate directly onto existing vegetation and land units. Differences in the spatial distribution of ecologically important climatic variables produce different patterns of potential impact. Variations in climatic conditions for different scenarios indicate that similar values of average change result in different ecosystem impacts across continental areas.

Potential impacts of global climate change on northern ecosystems must also consider existing patterns of human land use and resource management. Current fragmentation of vegetation cover will influence the response of ecosystems to changing climatic patterns and present barriers to long-term species migration (Peters and Darling 1985). Assessing migration barriers and resilience to climate change will require higher resolution spatial studies and dynamic ecosystem models (Halpin 1993).

Recent studies have demonstrated the sensitivity of terrestrial ecosystems to climatic disturbance at spatial scales ranging from site specific (e.g., Pastor and Post 1988; Bonan et al. 1990; Smith et al. 1992) to global (Emanuel et al. 1985; Prentice et al. 1992; Smith et al. 1992). In addition, paleoecological studies have shown the long-term response of forest distribution to past changes in global climatic patterns (e.g., Webb 1987; Davis 1989). This chapter investigates ecosystem impacts in the context of ongoing research in the field of climatic system dynamics (Chapter 2). Estimates of physiological responses of vegetation to climate change (Chapter 5), rates of species migration, and the effects of landscape fragmentation are speculative at this time. However, a qualitative assessment of risks to ecosystems from climate change may be made through cautious analysis of climatic scenarios.

Objectives and approach

The objectives of this chapter are to (1) estimate the impacts of climate change on high-latitude North American ecosystems and nature reserves, and (2) identify potential implications of changing climatic patterns on biotic resources in the Far North. Six northern biomes (characteristic vegetation formations) are used as a baseline for analysis: tundra, coniferous forest, deciduous forest, grass/shrublands, desert, and wetlands. Each of these biomes identifies a region with coarsely defined vegetation structure, species composition, and general climatic character-istics. This simplified approach allows for the discrimination of variation in the spatial distribution of climatic impacts.

Human intervention on the landscape must also be mapped to differentiate areas of natural vegetation from areas of fragmentation or agricultural use (Halpin 1993; McCloskey and Spalding 1989). In addition, policy and manage-ment activities in natural versus managed landscapes are typically different (Halpin and Secrett 1994). A digital database of global vegetation and land use is used to differentiate the land area of North America into three land regions: (1) predominantly agricultural, (2) agricultural/forest mosaic, and (3) significant areas of natural vegetation. Natural vegetation cover should not be confused with strict wilderness or roadless areas (McCloskey and Spalding 1989).

Most of northern North America is relatively free of landscape fragmenta-tion. However, proximity to current agricultural and forest areas must be consid-ered when assessing long-term climate change scenarios, because the amount and spatial extent of agriculture and forestry in northern regions could change under a different climate. Areas now unsuitable for agriculture may become more viable in future climates, and changes in the market for agricultural and forest products may make economically marginal lands more productive. Analysis of potential landscape changes and ecosystem impacts must consider both the direct ecologi-cal effects of climate change and the indirect effects of human land use.

General description of models

Two general methods were used in this assessment: (1) ecoclimatic models that correlate the distribution of vegetation with current climatic patterns, and (2) bio-physical models of moisture change (evapotranspiration).

Climate-vegetation classification

Climate-vegetation classification models relate vegetation pattern to climate at a global scale. These ecoclimatic models relate vegetation distribution or plant types (e.g., biomes) to biologically important features of the climate (e.g., tem-perature, precipitation) to define ecoclimatic zones (Thornthwaite 1948; Holdridge 1949; Prentice 1990; Prentice et al. 1992). Ecoclimatic zones therefore define the climatic conditions associated with a given vegetation type (e.g., boreal forest, tundra). Ecoclimatic classification models have been used to simulate vegetation distribution both for climatic conditions associated with the last glacial maximum (Manabe and Stouffer 1986; Hansen et al. 1988) and for predictions of future climatic patterns under increased CO_2 (Emanuel et al. 1985; Prentice et al. 1992; Smith et al. 1992).

Holdridge Life Zone ecoclimatic classification

The Holdridge Life Zone Classification (Holdridge 1967) is used in this analysis. Two variables, annual biotemperature and average annual precipitation, determine the classification. A third variable, potential evapotranspiration (the amount of water that would be released to the atmosphere under conditions of sufficient available water throughout the growing season), is estimated by the relationship between biotemperature and precipitation. Ecoclimatic "life zones," such as deciduous forest, boreal forest, or tundra, are defined by ranges of temperature and precipitation combinations—a kind of climatic "envelope." The complete Holdridge classification at a global scale includes 37 ecoclimatic life zones or biome types that can be divided into altitudinal zones and transitions at regional scales.

Although the Holdridge classification provides a description of the association between vegetation and regional climate, the model assumes that existing vegetation is in equilibrium with current climatic patterns. This presents a problem in the application of the models to climate change scenarios. The time scale of climatic changes (i.e., 70–100 years) may be faster than potential movement of vegetation. Therefore, this method provides a means of evaluating the *potential* for a geographical area to sustain a vegetation type, not whether that vegetation type will occur there following climate change. Only areas experiencing a significant amount of climate change are identified.

Models of evapotranspiration change

The use of changes in ecoclimatic zones as an indicator of potential long-term ecosystem stress assumes that a climatic shift may lead to changes in physiological conditions, vegetation composition, and interspecies competition. An index of effective moisture change is used to facilitate mapping the short-term direction of potential change. This mapping indicates short-term changes in climatic variables in the time frame of the GCMs (Manabe and Weatherald 1987; Kellogg and Zhao 1988).

Evapotranspiration is the total water lost to the atmosphere from evaporation from the land surface and transpiration from plants. This assessment uses a model that calculates evapotranspiration and is applied on a daily basis, based on daily temperature, solar irradiance, and precipitation values, to provide daily estimates of soil water deficits (Priestly and Taylor 1972); soil texture and available water capacity are from the Food and Agriculture Organization's soils database.

The Priestley-Taylor model is used to estimate potential (PET) and actual (AET) evapotranspiration on both a monthly and annual basis. PET represents the total moisture demand at a site; AET is the amount of water released to the atmosphere under the constraints of precipitation and soil water availability. The ratio AET/PET is an index of plant moisture stress, essentially a ratio of supply to demand. A high value of this scalar index (0 to 1) means that moisture supply meets or exceeds demand, while a low value means that moisture demand exceeds supply. This index identifies the combined effects of temperature and precipitation changes at a site. Absolute moisture deficit (D = PET − AET) (Stephenson 1990) is another important measure of potential ecosystem moisture change relevant to vegetation patterns. A deficit estimates the amount of additional moisture required for a site to meet PET demands.

Potential changes in ecoclimatic regions

The current distributions of Holdridge Life Zones were mapped using a standard climatic database of mean monthly precipitation and temperature at a $0.5° \times 0.5°$ (latitude and longitude) resolution (Leemans and Cramer 1990). Simulations of doubled CO_2 climates from four GCM scenarios (Table 4.1) were used to construct climate change scenarios by taking the difference between simulated current and doubled CO_2 climates. The change in climate from each GCM was then applied to the global climatic database. The altered databases corresponding to each of the four GCM scenarios were then reclassified using the Holdridge Life Zone classification and calculations of moisture change.

The four climate change scenarios used here are well-known models used in recent assessments of climate change impacts. Simulated climates depart from actual climatic data to a higher degree over western mountains and large bodies of water bodies such as Hudson Bay in northern North America. The GFDL model exhibits higher precipitation than expected in the Alaskan mountains and north-

Table 4.1. General circulation models used to construct climate change scenarios.

General circulation model	Temperature (°C)	Precipitation (%)
OSU—Oregon State University (Schlesinger and Zhao 1988)	+2.8	+7.8
GFDL—Geofluid Dynamics Laboratory (Manabe and Wetherald 1987)	+4.0	+8.7
GISS—Goddard Institute for Space Science (Hansen et al. 1988)	+4.2	+11.0
UKMO—United Kingdom Meteorological Office (Mitchell 1983)	+5.2	+15.0

central Canada, but drier conditions than expected along the Alaskan coast. The GISS model produces closer agreement for precipitation across high latitudes, with drier conditions along the Alaskan coast. The OSU scenario shows good agreement with current precipitation conditions across northern land areas, with drier than expected conditions along the Alaskan coast. With the exception of the Gulf Coast of Alaska, the UKMO scenario exhibits slightly more precipitation at high latitudes.

All four scenarios produce significant shifts in the distribution of northern ecoclimatic zones under doubled CO_2 climates (Figure 4.1; Table 4.2) (Emanuel et al. 1985; Smith et al. 1992). In this analysis, areas listed as stable did not receive sufficient climatic forcing to change the ecoclimatic classification. This does not mean that climatic changes are not predicted in those regions, but that the changes in those regions are within the range of the existing ecoclimatic class. Few Northern ecoclimatic regions remain stable under the UKMO and GFDL scenarios, while larger areas remain stable under the OSU and GISS scenarios (Figure 4.1).

The directions of change are similar for all GCM scenarios: The extent of tundra and polar desert climates decrease, while grassland and forest climates increase. These patterns are similar to those predicted in other North American studies (LeDrew 1986). The decline in tundra climates is primarily due to a shift from tundra to mesic (moderate temperature) boreal forest climates. This transition results from warming at higher latitudes and the subsequent northward movement of climates now associated with boreal forest regions into the areas now occupied by wet tundra.

Significant areas of the current boreal forest change to steppe or northern grassland climates. This trend is most widespread in the central continental regions of Canada. The southeastern portion of the current boreal forest region experiences notable shifts to climates now associated with cool temperate mixed-forest types under all scenarios. This trend is also observed for maritime regions of the Pacific Northwest and western Alaska.

Climates now associated with boreal forest tree line shift northward under all scenarios (Figure 4.1). While climates can change rapidly, any actual expansion of tree line into tundra would be limited by soil conditions, landscape hydrology, and seed dispersal. Shifts from tundra to forest at the northern edge of the current forest zone would likely follow a slowly moving front of young forest islands developing in front of the existing tree line. At the southern limit of the boreal forest, persistent tree species would more likely be replaced through the interaction of fire or windthrow and subsequent intrusion by more southerly plant species (Bonan et al. 1990).

Rates of vegetation change are difficult to predict because numerous environmental variables and disturbance events may interact to alter the landscape. In addition, difficulty in defining physiological tolerances of tree species hinders our understanding of ecosystem responses (Bonan and Sirois 1992). Paleoecolog-

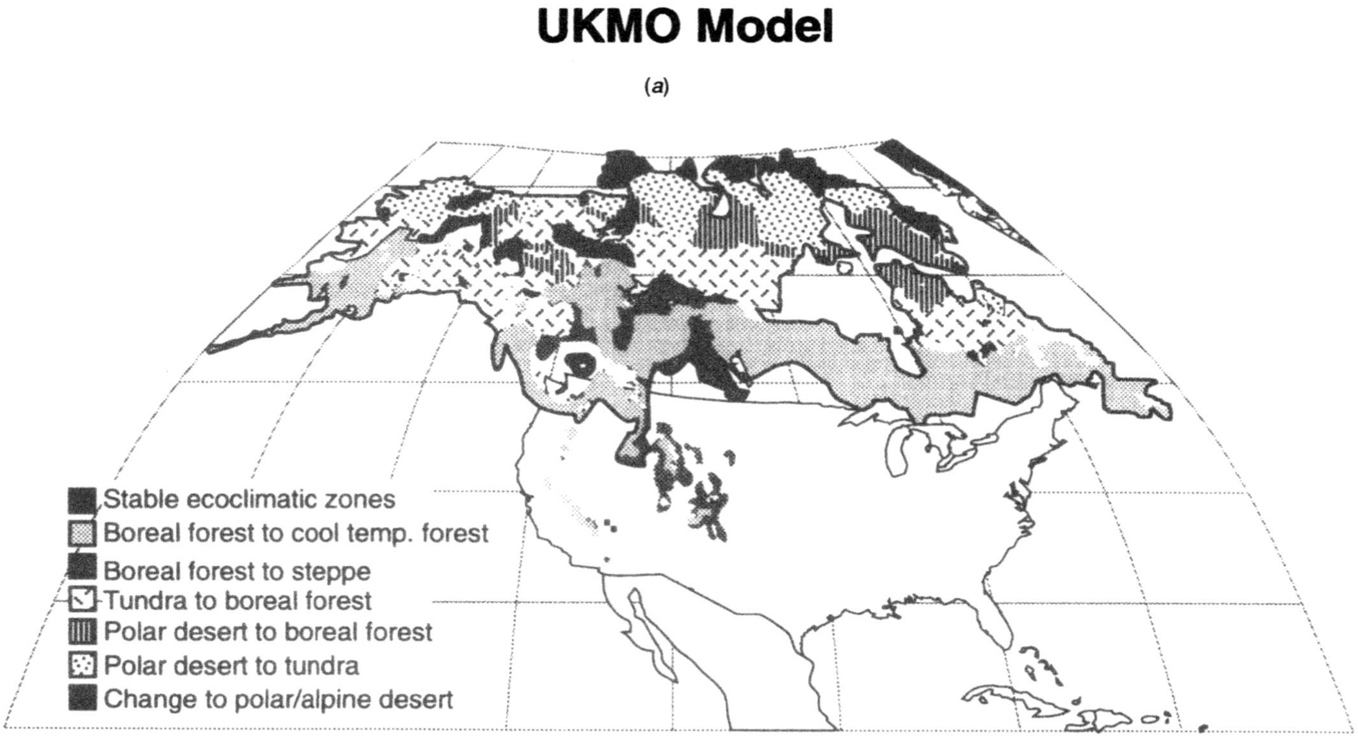

UKMO Model

(a)

Stable ecoclimatic zones
Boreal forest to cool temp. forest
Boreal forest to steppe
Tundra to boreal forest
Polar desert to boreal forest
Polar desert to tundra
Change to polar/alpine desert

Figure 4.1a–d. Spatial changes in ecoclimatic zones for major boreal and Arctic ecoclimatic zones under the UKMO, OSU, GFDL, and GISS climatic change scenarios.

OSU Model

(b)

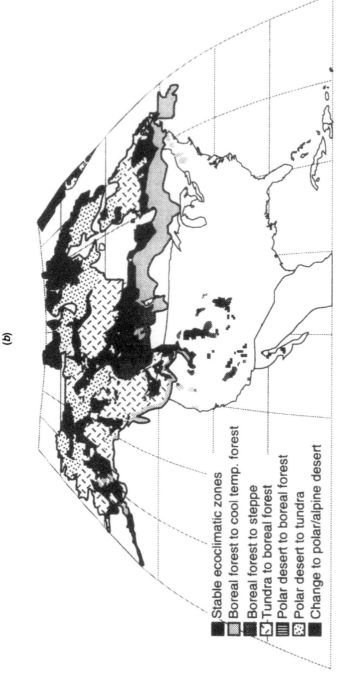

Stable ecoclimatic zones
Boreal forest to cool temp. forest
Boreal forest to steppe
Tundra to boreal forest
Polar desert to boreal forest
Polar desert to tundra
Change to polar/alpine desert

Figure 4.1b

GFDL Model

(c)

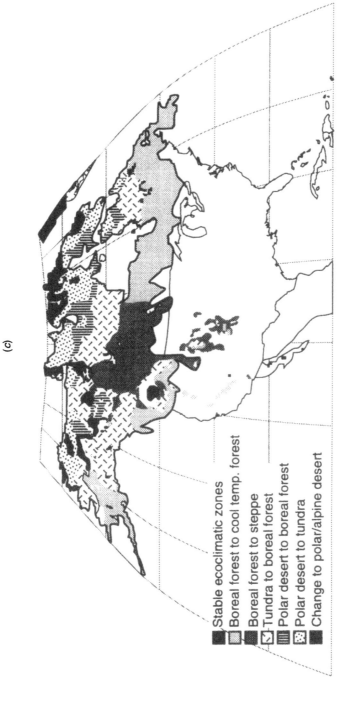

Stable ecoclimatic zones
Boreal forest to cool temp. forest
Boreal forest to steppe
Tundra to boreal forest
Polar desert to boreal forest
Polar desert to tundra
Change to polar/alpine desert

Figure 4.1c

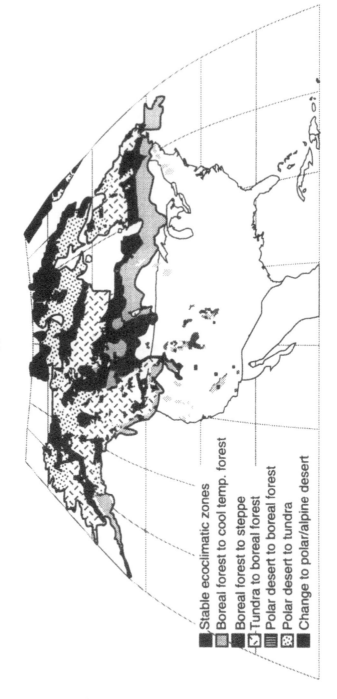

GISS Model

(d)

Stable ecoclimatic zones
Boreal forest to cool temp. forest
Boreal forest to steppe
Tundra to boreal forest
Polar desert to boreal forest
Polar desert to tundra
Change to polar/alpine desert

Figure 4.1d

Table 4.2. *Type of change in ecoclimatic zones as a percent of total area for scenarios generated by each GCM (see Table 4.1).*

	General circulation model			
Type of ecoclimatic change	UKMO	OSU	GFDL	GISS
Stable ecoclimatic areas	12.6	33.1	11.2	40.1
Boreal forest to cool temperate forest	33.6	14.8	23.6	19.7
Boreal forest to steppe	11.3	10.1	20.7	6.1
Tundra to boreal forest	19.8	17.8	21.3	19.3
Polar desert to boreal forest	8.0	4.7	7.2	0.5
Polar desert to tundra	12.9	13.6	12.6	13.1
Change to polar desert	1.8	5.8	3.3	1.2

ical data suggest that northern tree line movement may range from 10 to 45 kilometers per decade (Ritchie and MacDonald 1986; Davis 1989).

While the models have many similar outcomes, the four climate change scenarios produce different spatial patterns of change. A comparison of the areas of ecoclimatic change, and the ecoclimatic classes to which they changed, produces 17% agreement in area of change between the models globally (Leemans and Halpin 1992). Assertions of increased probability of change in areas of model overlap are inappropriate because the models represent independent simulation experiments with different techniques and interpretations of atmospheric dynamics.

Potential changes in effective moisture

The direction and magnitude of effective moisture change are shown in Figure 4.2. This assessment maps continuous change from areas of significant drying to areas of significant wetting and then overlays these changes onto existing vegetation type. The outcomes of simulated climatic changes in terms of actual vegetation distributions are presented instead of hypothetical new ecoclimatic zones. Change in effective moisture is an important factor in the alteration of hydrologic and ecophysiological response for a region (Chapter 5).

The direction of moisture index change (drier, no change, wetter) and a coarse classification of its magnitude into two levels of moisture change were assessed for each scenario. The four scenarios produce widely different moisture change outcomes and spatial patterns of change (Figure 4.2). The UKMO scenario produces significant drying in southeastern regions and moisture increases in southwestern regions and northern Alaska. Areas of change are mixed between effective wetting and drying trends for most major ecosystems with two obvious exceptions. Changes in moisture regime for existing deciduous forest and wetland areas are heavily skewed toward effectively drier climates. The majority of tundra and coniferous forest zones experience either an effective wetting or little

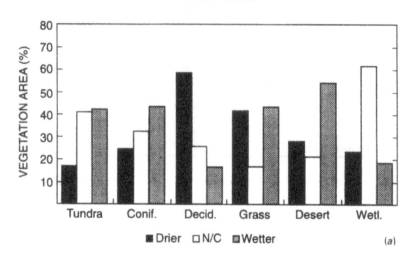

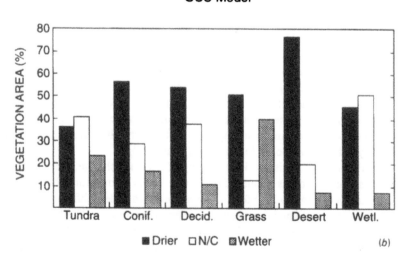

Figure 4.2a–b. Directional moisture ratio changes for major vegetation regions of North America (units in percentage area of each existing vegetation region). N/C indicates no change.

GFDL Model

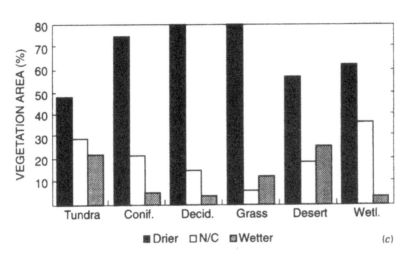

(c)

GISS Model

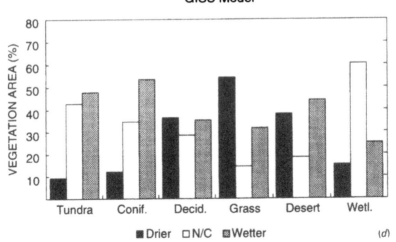

(d)

Figure 4.2c–d. Directional moisture ratio changes for major vegetation regions of North America (units in percentage area of each existing vegetation region).

change in moisture index. The OSU scenario produces relative drying across the continent with large patches of relative moisture increase in the Appalachians, north-central plains regions, and Alaskan mountain areas. Existing desert and wetland areas experience a significant trend toward effectively drier climates. Tundra and coniferous forest regions also experience significant drying trends.

The GFDL scenario provides the most extreme outcome for North America. With the exception of a portion of the southwestern states and northeastern Alaska, the GFDL scenario predicts a significant drying of the entire continent. Over 45% of existing tundra and 70% of coniferous forest change to drier climates under this scenario. The GISS scenario produces large areas of drying mixed with wetter patches in the southeastern and northern mountain regions. However, many northern regions experience relatively wetter conditions; existing tundra and conifer forest both exhibit this trend.

This analysis demonstrates that no single type of ecoclimatic change will occur across a continental-sized area. Simplistic presentations of climate change scenarios in terms of average warmer-drier or warmer-wetter conditions fail to show the wide distribution of outcomes presented in the scenarios. Under any given future scenario, different spatial distributions of existing vegetation may encounter different directions of climate change. Resource managers and planners attempting to anticipate changing environmental conditions will probably need to prepare for effective drying in one region and increased moisture in another.

Implications for northern habitats and protected areas

One important feature of global assessments of potential changes in ecoclimatic zones concerns the future condition of protected areas (Leemans and Halpin 1992). As a first assessment of the potential impacts of climate change on northern nature reserves, the locations of 169 United States and Canadian protected areas above 45° N latitude were entered into a digital database to assess the distribution of potential climate change impacts (Leemans and Halpin 1992; Halpin 1993). There are 13 biosphere reserve sites (designated by the UNESCO Man and the Biosphere Program) contained within these sampled protected areas. Biosphere reserves are designated within a regional management context, where the core natural areas are protected by buffer zones of multiple-use land management (Chapter 18). Modeling potential climatic impacts on these sites offers a perspective on risks to biosocial resources of an entire system.

Predicted ecoclimatic zone changes in Northern protected areas are 86% (UKMO), 46% (OSU), 76% (GFDL), and 59% (GISS). In fact, northern latitudes experience a higher rate of nature reserve impact than global averages (Leemans and Halpin 1992; Halpin 1994). This is due to the combination of higher levels of predicted climate change and higher numbers of established nature reserves in the Northern Hemisphere.

The four climatic scenarios clearly demonstrate that potential shifts in ecoclimatic zones and effective moisture classes could have a significant impact on the resources of individual reserves and on global reserve systems in general (Smith et al. 1990; Leemans and Halpin 1992). Predicted shifts in ecoclimatic zones are interpreted not as specific changes in vegetation type or species composition but as changes in climate associated with ecosystem types not currently represented by the reserve. Although these changes do not address individual species responses, the magnitude of change required to shift ecoclimatic zones would influence the suitability of habitats required by the component species. These changes could result in extinctions for species whose distributions are limited to the reserve system (Peters and Darling 1985; Graham 1988).

This analysis demonstrates that many northern parks and protected areas experience enough change in climate to reclassify an area from its current ecoclimatic classification (e.g., tundra to forest climate). Reserves are fixed sites set within a matrix of landscape features and land uses. As species potentially move to track changing climatic conditions, existing reserves may no longer contain the species assemblages they were intended to protect (Peters and Darling 1985). Expanding the size, shape, or pattern of nature reserves to new configurations in response to changing climates would require detailed knowledge of species response to changing environmental conditions. Reallocating land areas for protection would likely exacerbate conflicts over resource management and subsistence uses (Chapters 19 and 20). Maintenance of corridors for species movement in response to changing climatic patterns is a basic management issue that will increase in importance if habitat distribution is altered in the future.

Summary

This overview identifies areas of potential ecosystem change for high-latitude regions of North America at a continental scale. Spatial analysis indicates that a high percentage of existing natural ecosystems in the northern United States and Canada could be impacted by significant changes in climate under the scenarios tested. Expected changes in effective moisture demonstrate that significant percentages of land area would have altered moisture regimes in all ecosystem types. Spatial distributions of climatic changes derived from GCMs differ substantially. Although spatial variation in the modeled results is high, there is clearly a general trend toward effectively drier climates in most existing natural areas.

Simulated changes in ecoclimatic zones and effective moisture regimes for critical northern habitats and protected areas indicate that boreal forest and Arctic biological resources may be at risk from climatic changes. However, more detailed analysis of potential ecosystem responses is needed. Higher resolution analysis of the effects of changing climatic patterns on species assemblages, especially in critical protected areas, is needed to identify the sensitivity of specific sites to climatic perturbations. In addition, a better understanding is needed of the processes that control long-term change at the landscape level.

Translating potential climatic and ecological changes into meaningful indicators of future impacts on resource management and subsistence use is a challenging and uncertain process. However, early identification of potential ecological responses to climate change will allow for assessment of inherent risks, collection of baseline data for monitoring, and a wider range of future management options.

References

Bonan, G. B., H. H. Shugart, and D. L. Urban. 1990. The sensitivity of some high-latitude boreal forests to climatic parameters. *Climatic Change* 16:9–29.

Bonan, G. B., and L. Sirois. 1992. Air temperature, tree growth, and the northern and southern range limits of *Picea mariana*. *Journal of Vegetation Science* 3:495–506.

Chapin, F. S., R. L. Jeffries, J. F. Reynolds, G. R. Shaver, and J. Svoboda. 1992. *Arctic ecosystems in a changing climate*. New York: Academic Press.

Davis, M. B. 1989. Lags in vegetation response to greenhouse warming. *Climatic Change* 15:75–82.

Emanuel, W. R., H. H. Shugart, and M. P. Stevenson. 1985. Climatic change and the broad-scale distribution of terrestrial ecosystem complexes. *Climatic Change* 7:29–43.

Graham, R. W. 1988. The role of climate change in the design of biological reserves: The paleoecological perspective for conservation biology. *Conservation Biology* 2:391–394.

Halpin, P. N. 1993. *United States ecosystems at risk to climate change*. Report to the Congress of the United States. Washington, DC: Office of Technology Assessment.

Halpin, P. N. 1994. Latitudinal variation in the potential response of mountain ecosystems to climatic change. In *Mountain environments in changing climates*, edited by M. Beniston, 180–204. London: Routledge.

Halpin, P. N., and C. M. Secrett. 1994. Potential impacts of climate change on forest protection in the humid tropics: A case study in Costa Rica. In *Impacts of climate change on ecosystems and species*. Gland, Switzerland: IUCN (World Conservation Union). In press.

Hansen, J., I. Fung, A. Lacis, S. Lebedeff, D. Rind, R. Ruedy, G. Russel, and P. Stone. 1988. Global climate changes as forecast by the Goddard Institute for Space Studies three dimensional model. *Journal of Geophysical Research* 93:9341–9364.

Henderson-Sellars, A. 1990. Predicting generalized ecosystem groups with the NCAR GCM: First steps towards an interactive biosphere. *Journal of Climate* 3:917–940.

Holdridge, L. R. 1949. Determination of world plant formations from simple climate data. *Science* 105:367–368.

Holdridge, L. R. 1967. *Life zone ecology.* San Jose, Costa Rica: Tropical Science Center.

Kellogg, W. W., and Z. Zhao. 1988. Sensitivity of soil moisture to doubling of carbon dioxide in climate model experiments. Part I. North America. *Journal of Climate* 1:348–366.

LeDrew, E. F. 1986. Sensitivity of the Arctic climate: A factor in developing planning strategies for our arctic heritage. *Environmental Conservation* 13:215–228.

Leemans, R., and W. P. Cramer. 1990. The IIASA database for mean monthly values of temperature, precipitation and cloudiness on a global terrestrial grid. WP–90–41. Laxenburg, Austria: International Institute for Applied Systems Analysis.

Leemans, R., and P. N. Halpin. 1992. Biodiversity and global change. In *Biodiversity: Status of the earth's living resources*, edited by J. McComb, 254–255. Cambridge: World Conservation Monitoring Center.

Manabe, S., and R. J. Stouffer. 1986. Sensitivity of a global climate to an increase in CO_2 concentration in the atmosphere. *Journal of Geophysical Research* 85:5529–5554.

Manabe, S., and R. T. Wetherald. 1987. Large scale changes in soil wetness induced by an increase in carbon dioxide. *Journal of Atmospheric Sciences* 44:1211–1235.

McCloskey, J. M., and H. Spalding. 1989. A reconnaissance-level inventory of the amount of wilderness remaining in the world. *Ambio* 18:221–227.

Pastor, J., and W. M. Post. 1988. Response of northern forests to CO_2–induced climate change. *Nature* 334:55–57.

Peters, R. L., and J. D. Darling. 1985. The greenhouse effect and nature reserves. *Bioscience* 35:707–717.

Prentice, K. C. 1990. Bioclimatic distribution of vegetation for GCM studies. *Journal of Geophysical Research* 95:11811–11830.

Prentice, I. C., W. Cramer, S. P. Harrison, R. Leemans, R. A. Monserud, and A. Solomon. 1992. A global biome model based on plant physiology and dominance, soil properties and climate. *Journal of Biogeography* 19:117–134.

Priestley, C. H. B., and R. J. Taylor. 1972. On the assessment of surface heat flux and evaporation using large-scale parameters. *Monthly Weather Review* 100:81–92.

Ritchie, J. C., and G. M. MacDonald. 1986. The patterns of post-glacial spread of white spruce. *Journal of Biogeography* 13:527–540.

Schlesinger, M., and Z. Zhao. 1988. *Seasonal climatic changes induced by doubled CO_2 as simulated by the OSU atmospheric GCM/mixed layer ocean model.* Corvallis: Climate Research Institute, Oregon State University.

Smith, T. M., H. H. Shugart, G. B. Bonan, and J. B. Smith. 1992. Modeling the potential response of vegetation to global climate change. *Advances in Ecological Research* 22:93–113.

Smith, T. M., H. H. Shugart, and P. N. Halpin. 1990. Global forests. In *Progress reports on international studies of climate change impacts*, 25–32. Washington, DC: U.S. Environmental Protection Agency.

Solomon, A. M. 1992. The nature and distribution of past, present and future boreal forests: Lessons for a research and modeling agenda. In *A systems analysis of the global boreal forest*, edited by H. H. Shugart, R. Leemans, and G. B. Bonan, 291–307. Cambridge: Cambridge University Press.
Stephenson, N. L. 1990. The climatic control of vegetation distribution and the role of the water balance. *The American Naturalist* 135:649–670.
Thornthwaite, C. W. 1948. An approach toward a rational classification of climate. *Geographical Review* 38:55–89.
Webb, T. 1987. The appearance and disappearance of major vegetational assemblages: Long-term vegetation dynamics in eastern North America. *Vegetatio* 69:177–187.

5

Climate and ecological relationships in northern latitude ecosystems

John L. Hom

Boreal forest and Arctic tundra, the principal ecosystems of northern North America, have become a focal point of scientific, economic, and Native concerns with respect to the potential impacts of a warmer climate (Roots 1989). General circulation models predict that the circumpolar regions will have the greatest increases in mean annual air temperature with an expected doubling of atmospheric CO_2 concentration in the next century (Chapters 2 and 4). Temperatures are predicted to increase up to 2°C in summer and 10°C in winter for the polar regions; surface temperatures in interior Alaska are expected to increase 8° to 12°C in winter and up to 4°C in summer (see Table 2.1).

Northern ecosystems comprising the Arctic tundra and boreal forest (The terms boreal forest and taiga are used interchangeably to refer to northern latitude forest dominated by conifers and including hardwoods.) contain up to 455 gigatons (billion metric tons) of carbon in the active (upper nonpermafrost) soil layer and upper levels of permafrost (Gorham 1991); this amounts to 27% of the Earth's terrestrial carbon contained within 14% of its land area (Figure 5.1; Oechel and Vourlitus 1994). Warmer temperatures may make this region a net source of carbon dioxide (CO_2) flux due to increased soil respiration, and provide a strong positive feedback for an even greater greenhouse effect (Chapter 2).

These sensitive northern systems may serve as an early indicator that climate change is occurring, as well as provide records of past climatic fluctuations and their outcomes. Evidence for changes in northern climates and ecosystems are provided by recent investigations that have focused on the latitudinal tree line-tundra border, analysis of permafrost temperature profiles, tundra ecosystem CO_2 flux, and soil warming.

Physical environment

Permafrost, ice, and snow are dominant features in the Far North. Permafrost underlays 24% of the Earth's land surface (including mountains) and as much as

PLANT BIOMASS

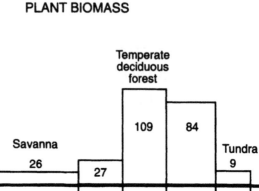

Figure 5.1. Carbon distribution among terrestrial ecosystems with respect to plant biomass and soil organic matter. The width of each bar represents the proportional land area of each ecosystem. Data are from Van Cleve and Powers (1995), figure 1. Reprinted with permission of American Society of Agronomy.

80% of Alaska (Figure 5.2). Permafrost is classified into two zones: continuous (mean average air temperature is less than $-6°C$) and discontinuous (mean average air temperature is greater than $-1°C$). In the continuous zone, permafrost occurs near the surface of the landscape, forming from the surface down. If winter ice development is greater than summer thaw, the thickness of the permafrost increases. At the southern limit, discontinuous isolated masses of permafrost are found under organic-rich soils, peat bogs, and north-facing slopes.

The predicted increase in mean annual air temperature would result in temperatures exceeding the threshold for maintaining permafrost and may result in a reduction of the vast discontinuous permafrost zone (Figure 5.2). Thawing of local areas of permafrost would liquify thawed soils, which would increase thermal erosion and freeze-thaw cycling. Models of permafrost dynamics indicate that the depth of thaw would increase 24% with a 6°C increase in temperature for Fairbanks, Alaska (mean annual air temperature of −3.5°C); surface energy balance and length of the snow-free season can moderate these temperature increases (Goodwin et al. 1985).

Additional factors that could ameliorate the predicted warming scenario for northern regions include cloud cover, snowfall, and precipitation. Surface energy budgets are modified greatly by the length of the snow-free season, total amount

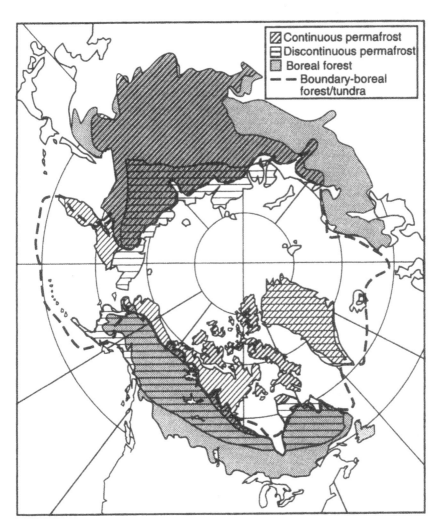

Figure 5.2. Circumpolar distribution of boreal forest, tundra, and permafrost. From Van Cleve and Dyrness (1983), figure 1. Reprinted with permission of Canadian Research Council.

of snow, albedo (percentage of incoming radiation that is reflected), moisture, cloudiness, and total radiation input. Increased snow accumulation may result in soil warming and altered distribution of permafrost. On the other hand, the increased snow depth could delay snowmelt and alter radiation with its high albedo. Increased precipitation would increase cloud cover, reducing radiation input to the surface. If summer precipitation decreases, increased radiation loads would result in warmer soils and increase the depth of active layers. However, some highly organic soils with low bulk density (soil weight per unit volume, typically in grams per cubic centimeter) may have lower heat transfer characteristics and act as better insulators.

Arctic and boreal ecosystems

Approximately 13% of the Earth's carbon is in the cold, permafrost-dominated soils of the boreal forest, and an additional 14% is sequestered in the cold soils of the tundra (Figure 5.1; Gorham 1991). Low soil temperature, low pH, low nutrient content, and anaerobic conditions in organic soils contribute to slow decomposition and accumulation of large quantities of carbon. In recent geological history, northern latitude ecosystems have been a major sink for atmospheric carbon, much of it accumulated in the past 10,000-year postglacial period (Post et al. 1982, Emanuel et al. 1985)

The Arctic is a treeless environment, characterized by a short summer growing season (6–14 weeks) with continuous daylight and a long winter with several months of continuous darkness. The low solar angle at these high latitudes reduces annual radiation input considerably. Although the 24-hour photoperiod during the peak Arctic summer is equivalent to the daily total radiation input in temperate regions, cloud cover during the summer can reduce incoming solar radiation by 50%. The vegetation of the Arctic is relatively low in diversity, but highly variable spatially. It is dominated by bryophytes (Bryophyta, typically mosses), sedges (Cyperaceae), tussock-forming grasses, low deciduous shrubs, and evergreen shrubs. Arctic vegetation typically has low stature and low annual carbon gain, with morphological adaptations to wind, winter desiccation, and ice abrasion. Plant species are well-adapted to short growing seasons and cold soils that limit nutrient uptake and carbon gain.

The boreal forest-tundra ecotone is distinguished by tree line, the northernmost limit of tree distribution, which coincides with the July 13°C isotherm, roughly the southern extent of the Arctic front during the summer. The boreal forest occupies the discontinuous permafrost zone. The boreal forest extends roughly north of 50° latitude (Figure 5.2). Its southern limit is less abrupt, occurring along the July 18°C isotherm, which is the average location of the Arctic front in winter. Only recently has the taiga been so extensive; it reached its maximum northward extent during the mid-Holocene (8,000–5,000 B.P.), occupying area that was once largely glaciated (Ritchie 1984).

Structure and function of the boreal forest

Many of the climatic, physical, chemical, and biotic factors that control ecological processes for growth, nutrient cycling, and physiological adaptation in the Far North can be inferred by examining the extensive literature on the boreal forest ecosystem (Larsen 1980; Van Cleve et al. 1983a; Van Cleve and Yarie 1986; Chapin et. al. 1992). The taiga forest in interior Alaska is a mosaic of deciduous and coniferous vegetation that reflects the influence of slope, aspect, drainage and other soil properties, local climate, and past fire disturbance (Viereck 1979). Periodic fires renew nutrient supplies and initiate a secondary succession cycle (temporal progression of different groups of plant species following a disturbance). Black spruce (*Picea mariana*) forests are the most extensive taiga vegetation type and occupy 44% of the forested area in Alaska. It is also the least productive forest type and is characteristically found in permafrost areas with the coldest, wettest soils. As black spruce begins to dominate the earlier successional stages of herbs and shrubs, the conifer canopy develops (40–60 years post-fire), creating favorable conditions for the rapid development of feathermosses on the forest floor (Van Cleve and Dyrness 1983).

With the establishment of mosses, a thick organic layer develops that progressively accumulates available nutrients, reduces soil temperature, and decreases the thickness of the active layer as depth of thaw decreases. Forest floor depths may reach 40 cm in older black spruce stands (Viereck et al. 1983). This serves as an insulating barrier that maintains cold soil temperatures. Mosses also compete with the vascular overstory for nutrients on these nutrient-deficient sites and act as filters that absorb atmospheric inputs of nutrients as well as litterfall (Oechel and Van Cleve 1986). Soil moisture increases as forest floor organic matter accumulates, the result of a shallow active layer over permafrost. These wetter conditions are favorable for further moss productivity, but may inhibit vascular plant growth due to acidic and anaerobic conditions in the rooting zone.

The mosses further restrict the chemical quality, or energy supply, of the forest floor material for decomposer activity. Black spruce sites have high forest floor lignin content and a high carbon:nitrogen ratio (44), twice that found in hardwood taiga sites. This indicates a strong potential for microbial immobilization of nitrogen and limited element recycling to the vascular overstory (Flanagan and Van Cleve 1983). Turnover of biomass (50 years) and macronutrients (61 years, for critical elements such as nitrogen, phosphorus, calcium, etc.) at black spruce sites is two to three times greater than for deciduous taiga species and an order of magnitude greater than for more productive temperate forests (Van Cleve et al. 1983b). The dynamics of forest floor organic matter indicate that mosses control soil temperature and quality of the organic substrate, thereby restricting microbial activity, decomposition, and nutrient cycling.

The black spruce ecosystem "conserves" nutrients more strongly as it ages, with slower rates of decomposition and nutrient turnover. This produces nutrient-limiting conditions for black spruce growth and reduces primary productivity. In older stands of black spruce, extreme nutrient deficiency can occur, with mosses producing more biomass annually than black spruce. With advancing succession, the nutrient capital of the ecosystem becomes progressively less available, standing dead material accumulates, and spruce productivity stagnates until the stand is again subjected to fire (Van Cleve and Viereck 1981).

Black spruce has adapted to low nutrient availability by increasing the efficiency with which it utilizes nutrients. Its evergreen growth form results in greater photosynthetic efficiency with respect to nitrogen and phosphorus compared to deciduous trees. The longevity of needles, with needles retained for up to 20 years, increases nutrient use efficiency and net carbon gain per unit nutrient (Hom and Oechel 1983).

Soil temperature as a control on ecosystem processes

Soil temperature is the dominant control of taiga ecosystem processes (Van Cleve et al. 1983a; Van Cleve and Yarie 1986). Slope and aspect are important to ecosystem structure and function because they control the amount of solar radiation received and initial soil temperature regimes. In later successional stages, biotic features such as vegetation cover, decomposition, and forest floor depth control soil temperature.

The highest annual production is found on the warmest sites: south-facing slopes with good drainage, where paper birch (*Betula papyrifera*) and white spruce (*Picea glauca*) dominate warmer soils (Van Cleve et al. 1983b; Viereck et al. 1983; Van Cleve and Yarie 1986). Although these sites are the most productive, they are nutrient-limited. Field studies indicate that nitrogen fertilization can increase annual aboveground biomass increment of aspen (*Populus tremuloides*) by 100% and increase annual diameter growth of white spruce by 100% (Van Cleve et al. 1983b).

The quality of forest floor organic matter interacts with soil temperature in controlling decomposition rates and nutrient cycling. As nutrients are increasingly retained in aboveground biomass and forest floor organic matter, litter becomes more resistant to breakdown. Large accumulations of organic matter on the forest floor result in lower forest floor and mineral soil temperatures and reduced decomposition rates.

Disturbance

Fire is the primary disturbance in upland boreal forests and may increase in frequency if climate becomes warmer and drier. Fire increases soil temperature by removing the overstory, reducing forest floor depth, and decreasing albedo, resulting in greater radiant heat transfer to the soil. The thickness of the active layer increases as warmer soil temperatures increase the depth of thaw. Warmer soil temperatures and greater resource availability stimulate microbial activity, which in turn increases decomposition and nutrient availability (Van Cleve and Viereck 1981).

Boreal forest burns on a 50–200 year cycle, with about 400,000 hectares burned annually in Alaska (Viereck 1973). Three million hectares of boreal forest in the former Soviet Union burned from 1980 to 1989, representing 38% of all forest fires for the Northern Hemisphere (Auclair 1993). Greater fire frequency during this period coincided with above average (0.5°–0.8°C) temperatures in the Northern Hemisphere. Fire has a positive effect on wildlife species such as moose (*Alces alces*) and snowshoe hare (*Lepus americanus*), which depend on early successional plant species for forage. Fire can have a negative effect on species such as caribou (*Rangifer tarandus*), which depend on lichens as a winter food source.

Physiological limitations at tree line

Tree line is often referred to as an indicator of climate change because it is the northernmost location at which environmental constraints prevent tree regeneration and growth. Physiological and reproductive processes that limit tree establishment are at the compensation point in terms of carbon gain and nutrient availability for tree species. Radiant energy captured at this high latitude determines length of growing season and depth of the active layer for nutrient uptake. Low soil temperatures not only reduce nutrient availability, but greatly reduce water uptake due to the higher viscosity of water at lower temperature and due to lower permeability of root membranes. This causes stomata (structures on leaf surfaces that open and close to control gas exchange) to close and limits carbon gain even when there is sufficient soil water and high humidity. A minimum soil temperature of 9°C in the active layer during summer is required for trees at their northern limit. Low temperature also affects the latitudinal limit of black spruce by affecting seed production and germination (Black and Bliss 1980).

Spruce are adapted to physiological limitations at tree line by having the ability to grow as stunted tree forms or upright forms. The stunted, or krummholz, form dominates where winter air temperature is very low, keeping foliage below the snow cover; reproduction is typically vegetative by layering (formation of roots and a new stem at locations where branch tips touch the soil). The upright form dominates under less harsh conditions; the apical buds (located at the branch tip) survive, enabling dispersal of seed for rapid invasion of suitable areas. Black spruce exists as the northernmost forest type by tolerating cold soils, low nutrient availability, and an active layer limited by permafrost and shallow depth of thaw.

Past changes and current observations in northern ecosystems

Northern ecosystems have been net sinks of carbon for the past 10,000 years. However, the apparent stability of that sink may be changing. Are northern ecosystems currently net sinks or sources of CO_2? What changes would occur with a warmer climate? Will the predicted changes in climate and other environmental conditions occur so quickly that these ecosystems will not be able to survive in their current locations?

During the Altithermal period (ca. 9,000–5,000 B.P.), temperatures were approximately 5°C warmer than they are now, and precipitation was higher. Alder (*Alnus* spp.) forests were found north of the Brooks Range in Alaska. The first boreal forests were established ca. 9,000 B.P. when white spruce spread rapidly from the Yukon region northward to the Brooks Range. Black spruce established ca. 6,000 B.P. and also became a dominant species. The modern distribution of both species was not established until ca. 4,000 B.P. Spruce migrations into the new habitat were relatively rapid—about 200 kilometers per century—an order of magnitude faster than has been documented for species such as American beech (*Fagus grandifolia*) (Davis 1989). The boreal forest also spread rapidly at its northern edge in Canada during the Altithermal period (MacDonald et al. 1993). Warm climatic intervals as short as 10 years have resulted in successful expansion of the spruce tree line in Canada and central Alaska (Viereck 1979).

Recent evidence from permafrost temperature profile records shows that Arctic regions may have warmed 2°–4°C during this century, with much of this warming occurring since 1960 (Lachenbruch and Marshall 1986). Geothermal reconstruction of the continuous permafrost record shows a warming taking place in the upper permafrost (0.2–2.0 meters beneath the ground surface) of the Alaskan Arctic. The permafrost record represents a long-term, integrated measurement and therefore provides a unique fingerprint for the current warming trend in northern regions.

Recent studies of tundra vegetation provide further insight on the interaction of atmospheric CO_2 and temperature with Arctic ecosystems. Intact cores of coastal wet tundra plants and soil extracted during the winter were transferred to controlled environments with elevated CO_2 and temperature environments (Billings and Peterson 1992). Temperature increases of 4°–8°C changed the tundra from a net sink for atmospheric CO_2 to a source of CO_2. Lowering the water table and exposing the soil to aerobic conditions increased CO_2 efflux from the tundra to the atmosphere. Apparently CO_2 was not the primary factor limiting production, although the indirect effects of CO_2 and temperature altered the water table, decomposition, length of growing season, and nutrient availability.

Oechel et al. (1993) found that long-term in situ exposure of Arctic tussock tundra to elevated CO_2 did not lead to long-term increases in carbon storage. Rather, the tussock tundra adjusted to higher levels of CO_2 by lowering the photosynthesic rate; photosynthesis that was initially stimulated under high CO_2 eventually returned to the same rates as the ambient (nonelevated) CO_2 treatment in a matter of weeks. In contrast, tussock tundra grown under elevated CO_2 and 4°C higher temperature had increased photosynthesis and was a net sink for atmospheric CO_2 over a 3-year period; nutrient availability increased as a result of increased, temperature-stimulated soil mineralization (breakdown of organic soil constituents to release nutrients) rates (Tissue and Oechel 1987; Grulke et al. 1990; Oechel et al. 1993). The projected change from carbon sink to source in these ecosystems is attributed to (1) lowering of the water table by warmer conditions, (2) increased decomposition due to higher temperatures, (3) increased mineralization and nutrient availability, and (4) increased capacity of plants to utilize elevated levels of CO_2.

Data on carbon accumulation along a latitudinal transect from tussock to coastal tundra in northern Alaska (a possible analog for future change in the Arctic) suggest that high latitude ecosystems will continue to act as a small sink for atmospheric CO_2 with increased temperature, thus moderating global warming (Marion and Oechel 1993). However, recent measurements of short-term (decades to centuries) carbon flux show that these ecosystems are currently a net source of atmospheric CO_2 and may represent a strong positive feedback to global warming (Oechel et al. 1993; Oechel and Vourlitis 1994).

Soil warming studies have been used to test the hypothesis that cold soils control nutrient cycling and productivity of the black spruce forest system. When permafrost-dominated soil was heated $8°–10°C$ above ambient temperature for three growing seasons, the soil heat sum increased by 100%. This resulted in significant changes in forest floor decomposition (20% biomass decrease) and increased nutrient (available forms of nitrogen and phosphorus) availability (Van Cleve et al. 1983b, 1990). There were also substantial increases in spruce needle nutrient content, photosynthetic rate, and annual tree growth (Table 5.1; Hom 1986; Van Cleve et al. 1990). Despite the apparent effects of soil warming, the relatively low absolute increases suggest that black spruce may have a (genetically) limited capacity to utilize increased resources.

Summary: The future of northern ecosystems

Predicting how northern vegetation will respond to increased air temperature and atmospheric CO_2 is difficult because of uncertainty about the magnitude and rate of environmental changes. A wide range of feedbacks to biotic and biophysical processes must be considered, including decomposition, fire frequency, soil and litter quality, precipitation, and cloud cover (Table 5.2; Chapin et al. 1992).

If future climates result in warmer, drier soils and more frequent fires, early successional stages of spruce forests will become more common and mature stands less common. Viereck and Van Cleve (1984) suggest that a warmer, drier climate in interior Alaska would result in an expansion of aspen mixed with steppe-like vegetation (relatively treeless plain) on dry sites and paper birch on wet sites previously occupied by black spruce. Spruce at tree line could expand into tundra areas, resulting in an overall increase in northern forests. Forest productivity would increase due to more rapid turnover and availability of nutrients.

Table 5.1. Summary of the impact of soil warming on a black spruce forest.

	Control	Heated (+9°C)	Change (%)
Depth of thaw (cm)	57	115	+100
Soil degree days at 10 cm			
(May 20–Sept. 10)	563	1589	+180
Forest floor biomass (g/m²)	8630	6904	−20
Forest floor depth (cm)	24	7.5*	−70
Radial growth (mm/yr)	0.25	0.33	+33
Annual tree production (g/m²/yr)	94	125	+33
Maximum photosynthesis			
(mg CO₂/g/yr)	3.21	3.93	+22
Dark respiration (mg CO₂/g/hr)	−0.77	−0.87	+13
Needle nitrogen (%)	0.98	1.22	+25
Needle phosphorus (%)	0.082	0.142	+73
Soil			
pH	5.87	4.86	−1.0 pH
nitrogen (%)	0.96	1.52	
lignin (%)	19.6	15.1	
moisture (%)	178	100	
Net carbon change (g/m²)	+100–200	−1700	−775
	per year	(2 years)	g/m²/yr

Sources: Data compiled from Hom (1986) and Van Cleve et al. (1990).
Notes: Asterisk indicates that this is a calculated equivalent, not actual.

Emanuel et al. (1985) suggest that a transition from dry boreal forests to steppe-like vegetation with a doubling of atmospheric CO_2 would decrease boreal forests by 37% and tundra by 32%.

Paleoecological evidence indicates that northward shifts of plant species in response to warmer, drier climates have usually increased carbon accumulation in those regions. However, tree species migrations require decades to centuries, and it is likely that more carbon will be lost from northern ecosystems before changes in species composition can counteract these losses (Marion and Oechel 1993; Oechel and Vourlitis 1994).

The rate of future environmental changes will be more critical than their magnitude. The boreal forest expanded during the most recent postglacial period, on a temporal scale of a few thousand years. Future climate change of about the same magnitude is expected to occur in 100 years or less. A migration rate of 400–600 kilometers per 100 years would be required to track isotherms (lines indicating constant mean temperature, in this case the temperature associated with the northernmost extent of a species) as they move northward. This would be up to 10 times more rapid than known tree migration rates (Davis 1989; Mac-Donald et al. 1993) and may exceed the ability of some long-lived species to adapt, resulting in local extinctions.

Of all the uncertainties associated with future impacts of climate change on northern ecosystems, the most critical are potential biotic feedbacks that will

Table 5.2. *Certainty of climate change and its direct effects on vegetation, and the antici-pated impact of these changes on the function of tundra ecosystems.*

	Certainty of change	Impact of change
Change in environment	High	Low
Increased CO_2	High	Low
Increased air temperature	Low	High
Increased season length	High	Moderate
Increased cloudiness	Low	Low
Increased precipitation	Low	High
Increased soil temperature	Moderate	High
Increased thaw depth	Moderate	High
Increased drainage	Moderate	High
Increased nutrient availability	Low	High
Increased fire	Low	High
Decreased soil moisture	Low	High
Direct effects on vegetation		
Increased photosynthesis	Moderate	Moderate
Increased nutrient uptake	Low	High
Increased nitrogen fixation	Moderate	Moderate
Increased transpiration	Moderate	Moderate
Increased methane transport	Low	High
Increased production	Low	High
Increased nutrient status	Low	High
Indirect effects on ecosystems		
Increased litter quality	Low	High
Increased decomposition	Low	High
Increased herbivory	Low	High
Increased methane oxidation	Low	High
Increased fire frequency	Low	High
Decreased nutrient immobilization	Low	High
Feedbacks to Earth's atmosphere		
Increased CO_2 release	Low	High
Decreased albedo	Low	High
Decreased methane release	Low	Low

Source: From Chapin et al. (1991). Reprinted with permission of Academic Press.

affect soil and litter quality, decomposition rates, forest floor depth, permafrost depth, soil temperature, soil moisture, and fire frequencies. Arctic tundra and boreal forests are expected to face the largest and earliest effects of climate change. These huge biological storehouses of carbon will, in turn, provide feed-backs to the global atmospheric system.

References

Auclair, A. N. D. 1993. Forest wildfire as a recent source of CO_2 at northern lati-tudes. *Canadian Journal of Forest Research* 23:1528–1536.

Billings, W. D., and K. M. Peterson. 1992. Some possible effects of climatic warming on arctic tundra ecosystems of the Alaskan North Slope. In *Global warming and biological diversity*, edited by R. L. Peters and T. E. Lovejoy, 233–243. New Haven, CT: Yale University Press.

Black, R. A., and L. C. Bliss. 1980. Reproductive ecology of *Picea mariana* (Mill.) B.S.P., at tree line near Inuvik, Northwest Territories, Canada. *Ecological Monographs* 50:331–354.

Chapin, F. S., R. L. Jefferies, J. F. Reynolds, G. R. Shaver, and J. Svoboda. 1992. Arctic plant physiological ecology in an ecosystem context. In *Arctic ecosystems in a changing climate: An ecophysiological perspective*, edited by F. S. Chapin, R. L. Jefferies, J. F. Reynolds, G. R. Shaver, and J. Svoboda, 441–451. San Diego: Academic Press.

Davis, M. B. 1989. Lags in vegetation response to greenhouse warming. *Climatic Change* 15:75–82.

Emanuel, W. R., J. H. Shugart, and M. P. Stevenson. 1985. Climatic change and the broad-scale distribution of terrestrial ecosystem complexes. *Climatic Change* 7:29–43.

Flanagan, P. W., and K. Van Cleve. 1983. Nutrient cycling in relation to decomposition and organic-matter quality in taiga ecosystems. *Canadian Journal of Forest Research* 13:795–817.

Goodwin, C. W., J. Brown, and S. I. Outcalt. 1985. Potential responses of permafrost to climatic warming. In *The potential effects of carbon dioxide-induced climatic changes in Alaska*, edited by J. H. McBeath, 92–105. Miscellaneous Publication 83–1. Fairbanks: University of Alaska.

Gorham, E. 1991. Northern peatlands: Role in the carbon cycle and probable responses to climatic warming. *Ecological Applications* 1:182–195

Grulke, N. E., G. H. Riechers, W. C. Oechel, U. Helm, and C. Jaeger. 1990. Carbon balance in tussock tundra under ambient and elevated atmospheric CO_2. *Oecologia* 83:485–494.

Hom, J. L. 1986. Investigations into some of the major controls on the productivity of a black spruce (*Picea mariana* (Mill) B.S.P.) forest ecosystem in the interior of Alaska. Ph.D. dissertation. Fairbanks: University of Alaska.

Hom, J. L., and W. C. Oechel. 1983. The photosynthetic capacity, nutrient content, and nutrient use efficiency of different needle age-classes of black spruce (*Picea mariana*) found in interior Alaska. *Canadian Journal of Forest Research* 13:834–839.

Lachenbruch, A. H., and B. V. Marshall. 1986. Changing climate: Geothermal evidence from permafrost in the Alaskan Arctic. *Science* 234:689–696.

Larsen, J. A. 1980. *The boreal ecosystem*. Academic Press: New York.

MacDonald, G. M., T. W. D. Edwards, K. A. Moser, R. Pienitz, and J. P. Smol. 1993. Rapid response of treeline vegetation and lakes to past climate warming. *Nature* 361:243–246.

Marion, G. M., and W. C. Oechel. 1993. Mid-to late-Holocene carbon balance in arctic Alaska and its implications for future global warming. *Holocene* 3:193–200.

Oechel, W. C., and G. L. Vourlitis. 1994. The effects of climate change on land-atmosphere feedbacks in arctic tundra regions. *Tree* 9:324–329.

Oechel, W. C., and K. Van Cleve. 1986. The role of bryophytes in nutrient cycling in the taiga. In *Forest ecosystems in the Alaskan taiga: a synthesis of structure and function,* edited by K. Van Cleve, F. S. Chapin, P. W. Flanagan, L. A. Viereck, and C. T. Dyrness, 121–137. New York: Springer-Verlag.

Oechel, W. C., S. J. Hastings, G. Vourlitis, M. Jenkins, G. Riechers, and N. Grulke. 1993. Recent change of Arctic tundra ecosystems from a net carbon dioxide sink to a source. *Nature* 361:520–523.

Post, W. M., W. R. Emanuel, P. J. Zinke, and A. G. Stangenberger. 1982. Soil carbon pools and world life zones. *Nature* 298:156–159.

Ritchie, J. C. 1984. *Past and present vegetation of the far northwest of Canada.* Toronto: University of Toronto Press.

Roots, E. F. 1989. Climate change: High-latitude regions. *Climatic Change* 15:233–253.

Tissue, D. T., and W. C. Oechel. 1987. Response of *Eriophorum vaginatum* to elevated CO_2 and temperature in the Alaskan arctic tundra. *Ecology* 68:401–410.

Van Cleve, K., and C. T. Dyrness. 1983. Introduction and overview of a multidisciplinary research project: The structure and function of a black spruce (*Picea mariana*) forest in relation to other fire affected taiga ecosystems. *Canadian Journal of Forest Research* 13: 695–702.

Van Cleve, K., and J. Yarie. 1986. Interaction of temperature, moisture, and soil chemistry in controlling nutrient cycling and ecosystem development in the taiga of Alaska. In *Forest ecosystems in the Alaskan taiga: A synthesis of structure and function,* edited by K. Van Cleve, F. S. Chapin, P. W. Flanagan, L. A. Viereck, and C. T. Dyrness, 160–189. New York: Springer-Verlag.

Van Cleve, K., and L. Viereck. 1981. Forest succession in relation to nutrient cycling in the boreal forest of Alaska. In *Forest succession: Concepts and application,* edited by D. C. West, H. H. Shugart, and D. B. Botkin, 185–211. New York: Springer-Verlag.

Van Cleve, K., and R. F. Powers. 1995. Soil carbon, soil formation and ecosystem development. In *Carbon forms and functions in forest soils,* edited by W. W. McFee and J. M. Kelly, 155–200. Madison, WI: Soil Science Society of America.

Van Cleve, K. C., C. T. Dyrness, L. A. Viereck, J. Fox, F. S. Chapin, and W. C. Oechel. 1983a. Taiga ecosystems in interior Alaska. *BioScience* 33:39–44.

Van Cleve, K. C., L. K. Oliver, R. Schlentner, L. A. Viereck, and C. T. Dyrness. 1983b. Productivity and nutrient cycling in taiga forest ecosystems. *Canadian Journal of Forest Research* 13: 747–766.

Van Cleve, K., W. C. Oechel, and J. L. Hom. 1990. Response of black spruce (*Picea mariana*) ecosystems to soil temperature modification in interior Alaska. *Canadian Journal of Forest Research* 20:1530–1535.

Viereck, L. A. 1973. Wildfire in the taiga of Alaska. *Quaternary Research* 3:465–495.

Viereck, L. A. 1979. Characteristics of treeline plant communities in Alaska. *Holarctic Ecology* 2:228–238.

Viereck, L. A., and K. Van Cleve. 1984. Some aspects of vegetation and temperature relationships in the Alaskan taiga. In *The potential effects of carbon dioxide-induced climatic changes in Alaska*, edited by J. H. McBeath, 129–142. Miscellaneous Publication 83–1. Fairbanks: University of Alaska.

Viereck, L. A., C. T. Dyrness, K. Van Cleve, and M. J. Foote. 1983. Vegetation, soils, and forest productivity in selected forest types in interior Alaska. *Canadian Journal of Forest Research* 13:702–720.

6

Responses of Arctic ungulates to climate change

Anne Gunn

A warmer Arctic will not be beneficial to many populations of caribou (*Rangifer tarandus*), musk-oxen (*Ovibos moschatus*), and the people (*Homo sapiens*) who depend on them during the initial decades of climate change. Sweeping statements are risky, but this chapter gives the logic behind this one while acknowledging uncertainties in our understanding. The scale of the predictions is decades, covering the time of transition when changes in weather may be rapid. Longer-term changes occurring as plant communities adapt, such as a northward movement of the tree line, are not discussed here.

This chapter focuses on Arctic ungulates (caribou, musk-oxen) and people in Canada's Northwest Territories (NWT, area of 330 million hectares). It is assumed that the NWT is representative of other circumpolar regions at least in the context of Arctic ungulate ecology and climate change. The caribou in this region are Peary caribou (*R. t. pearyi*) on the Arctic islands and barren-ground caribou (*R. t. groenlandicus*) on the mainland and Baffin Island (Figure 6.1). The aboriginal people of this region are the Dene (Indians) and Métis, who mostly hunt in the taiga, or boreal forest (These terms are used interchangeably to refer to northern latitude forest dominated by conifers and including hardwoods.). Above the tree line are Inuit in the eastern Arctic and Inuvialuit in the western Arctic. Habitats range from the high Arctic deserts across treeless barrens to the taiga. Climate is Arctic continental and maritime with pronounced regional differences (Maxwell 1980).

People, caribou, and musk-oxen have been through climatic changes over at least the last 8,000 years—the span of human occupancy (McGhee 1990). People abandoned and then reoccupied areas as the climate changed, and they adapted their culture as the environment varied (Chapters 9 and 11). The relationships among climate, ungulates, and humans are explored in this chapter by (1) describing people's dependence on caribou and musk-oxen, (2) contrasting caribou and musk-ox ecology in relation to weather, (3) summarizing predictions of climate

Figure 6.1. Barren-ground caribou in autumn. Photo by Anne Gunn.

change for the Arctic, (4) evaluating information on the cumulative effects of weather on ungulate population size, and (5) predicting changes in caribou and musk-ox populations and what these mean to people in the NWT.

People, caribou, and musk-oxen

The NWT's economy is a mix of harvesting wildlife and wage earning. The 1991 population was 35,300 aboriginal and 22,300 nonaboriginal individuals (Bureau of Statistics 1994). Many still rely on hunting and fishing for food and income: over 90% of aboriginal households use meat taken by hunting or fishing. Wage earning generates the cash needed to buy equipment and supplies for hunting and fishing. Most hunting is for food; the replacement cost of wild food would be $37 million (Canadian dollars) if that food became unavailable. Horns, antlers, and skins are sold or manufactured into goods, and small operators commercially butcher and process meat from caribou and musk-oxen for local sale.

Aboriginal people can hunt caribou for subsistence use, but commercial use (sale of meat or any parts) is restricted by quotas. Quotas are allocated only if the subsistence use is less than the sustained harvest. Annual levels of harvest depend on where the caribou have moved relative to the communities; most hunting is in winter, and caribou rotate their winter ranges every few years. The 1990–91 harvest was a subsistence take of about 55,000 caribou, with 4,485 commercial tags available (Government of NWT, unpublished information). Musk-ox harvesting is regulated under a quota based on the trend in numbers. The quota is assigned to a community, and they decide whether the use is subsistence or commercial. In 1990–91, subsistence use was 442 and commercial use 160. In 1994, there were 7,160 available tags. These figures do not convey the importance of hunting. The value of caribou and musk-oxen far exceeds any dollar value. Animals and hunting are an integral part of aboriginal culture (Figure 6.2) and have been for some 8,000 years. Nonaboriginals also have cultural values for hunting, viewing, and simply knowing that the animals exist.

Weather and forage: Caribou and musk-oxen

Weather affects forage and therefore herbivore physical condition. Physical condition affects births and deaths, which drive changes in population size. These interactions explain the mechanisms of how weather can affect population size.

Weather and plant growth

The relationship between plant growth and weather is easily observed for some plants (Chapter 5). For example, the growth of Arctic white heather (*Cassiope tetragona*) depends on May precipitation and July temperature (Callaghan et al. 1989). Many Arctic plants are "conservative," completing their annual growth cycle in a relatively fixed and short period, but weather strongly affects some growth rates and the onset of growth (Svoboda 1977).

Figure 6.2. Inuit loading harvested caribou onto a sled, Pelly Bay, NWT. Photo by Ted Leighton.

In hummocky sedge-moss meadows, peak aboveground live biomass varies little between a cool and a mild season (Muc 1977). However, summer mean temperatures can affect flowering the following year. The flowers of cottongrass (*Eriophorum* spp.) on the calving ground of the Porcupine caribou herd increased from 13.6 flowers/m² in 1980 (1979 was a cool, wet summer) to 48.9 flowers/m² in 1981 (1980 was a warm, dry summer) (Russell et al. 1993).

Weather, forage, and caribou

Weather is closely linked to caribou and musk-ox body condition (fat and muscle reserves), which, in turn, is reflected in individual survival and raising of calves. Weather alters forage availability, which drives caribou and musk-oxen seasonal body condition. Summer forage through plant growth (absolute availability) is mediated by snow and ice in winter (relative availability). Weather during the previous plant growing season influences forage availability in winter. For example, lichens tend to grow on drier sites and have less productivity in dry summers. Those sites are often exposed with shallow snow cover where caribou preferentially feed. The energy costs of cratering (digging through snow) are partly borne by body reserves, but 75% of the energy used comes from foraging (Tyler 1987).

Although caribou are energetically efficient at cratering to reach forage, their efforts are 30% greater than foraging on bare ground. The cost in energy increases exponentially with sinking depth, and dense or crusted snow drains even more energy. A crust on the snow that would almost support a caribou's weight before collapsing raises the cost of walking by 570% (Fancy and White 1985). Deep snow and crusted snow also shift the relationship between wolves (*Canis lupus*) and caribou: If the caribou break through and flounder, then wolves have the advantage if they can travel on the snow's surface.

A cow's body reserves determine when its calf will be born and what its birth weight will be. Gestation is prolonged if a cow is in poor condition, and undersized newborn calves have a poorer chance of survival (Eloranta and Nieminen 1986). Barren-ground caribou calve either just before or during the plants' burst of new growth (green up). Cows bear the cost of lactation more from forage intake than from their body reserves (Parker et al. 1990). Timing of forage availability is a key element in the growth of calves. Caribou switch from one plant species to another as plants flower or leaf out; new growth is high in nutrients. Timing is largely determined by spring weather, while flowering quantity lags one year in response to the previous summer temperatures. Besides supporting a calf, a cow during the summer must reach a threshold level of condition to conceive during the October rut (Thomas 1982) as well as to survive the winter.

Biting insects are an affliction for caribou in summer. Caribou lose foraging time and expend energy trying to escape the depredations of mosquitoes (Culicidae) and warble flies (*Hypoderma tarandi*). Summer temperatures and windspeed largely determine insect activity (Russell et al. 1993). If the harassment is severe, the caribou lose condition. For example, warm summers (and more insects) are correlated with a decline in the weight of calves in the fall (Helle and Tarvainen 1984); lighter calves are less likely to survive winter if the snow is deeper than average (Eloranta and Nieminen 1986).

The timing of freeze-up is another link between weather and caribou ecology. A later freeze-up can cause deaths as caribou break through thin ice when they attempt to travel across lakes rather than detour through the bush (Miller and Gunn 1986). Crossing sea ice poses similar perils for caribou migrating between Arctic islands.

Weather, forage, and musk-oxen

The reproductive strategies of caribou and musk-oxen differ in timing and allocation of body reserves to reproduction. Musk-ox cows maintain their body reserves in summer at the expense of milk quality (White et al. 1989; Parker et al. 1990). Slower calf growth is compensated by later weaning than caribou, and cows may suckle calves until mid- or late winter (Parker et al. 1990). Musk-oxen calve during late April and early May, and peak milk production is two to three weeks later. On the Arctic islands and the eastern mainland, cows metabolize body fat to support lactation for about four weeks because greening vegetation will not be available until early June (Figure 6.3; White et al. 1989).

Musk-ox cows replenish their body condition during and after the rut even while lactating. That does not rule out the role of body condition in determining conception rates as occurs in other ungulates (Thomas 1982; White et al. 1989). One effect of the musk-oxen's early rut (August) is that they may be more responsive to summer conditions than caribou. They are more likely to invest in adult

Figure 6.3. Musk-oxen in early June, Victoria Island, NWT. Photo by Anne Gunn.

and calf survival at the expense of reproduction, but are less sensitive to winter conditions because of larger body size, which enables them to "coast" during the winter.

Musk-ox diet is dominated by willows (*Salix* spp.), sedges (Cyperaceae), and grasses (Poaceae). Musk-oxen select forbs, but that preference does not show up in analyses of fecal plant fragments, because forbs and the flowers of dwarf shrubs such as Arctic avens (*Dryas* spp.) and purple saxifrage (*Saxifraga opposi- tifolia*) are highly digestible (White et al. 1981). Forbs generally reach full size two to three weeks before grasses (Muc 1977).

Climate change predictions

A warming trend is expected during the next century, but the extent and rate of change are uncertain (Chapter 2). Maxwell's (1992) predictions for the eastern Canadian Arctic suggest that winters will be shorter, warmer, and wetter; snow- fall may increase as much as 30%, and freeze-thaw cycles will be more frequent. Summers will be warmer and wetter with increases in temperature and precipita- tion comparable to those increases projected for winter (Figure 6.4).

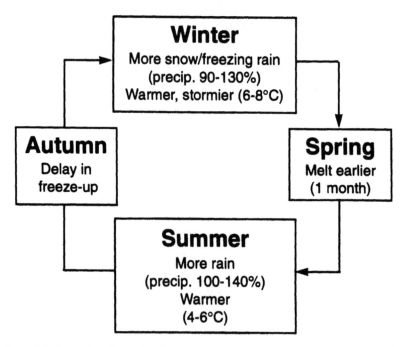

Figure 6.4. Seasonal predictions for climatic warming of mainland Arctic Canada. Adapted from data in Maxwell (1992). Reprinted with permission of Academic Press.

These climate change predictions are given as means, but their variability will almost certainly be high. In their analysis of weather data and musk-ox populations, Forchhammer and Boertmann (1993) commented that variability of winter weather was not related to mean temperatures. Variability and unpredictability of those changes may be the key to how ungulate populations respond.

Weather and fluctuations in caribou and musk-ox numbers

Fluctuations in caribou populations

Fluctuations in caribou numbers across the Arctic occur every few decades. In the Northwest Territories, the size of Qamanirjuaq, Beverly, and Bathurst herds of barren-ground caribou have changed by a factor of nearly 10 this century (Figure 6.5). The Qamanirjuaq herd reputedly numbered 100,000 before the 1950s, but declined to 33,000 in the late 1970s. It increased in the 1980s to 148,000–292,000

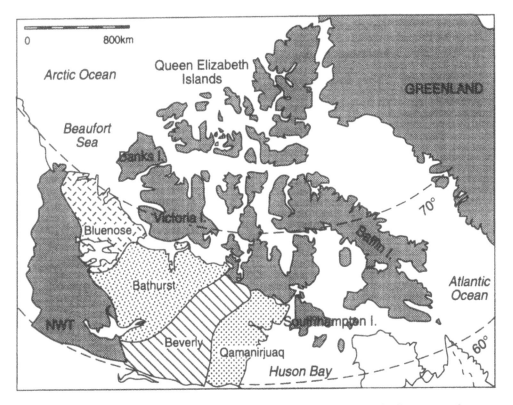

Figure 6.5. Arctic place names used in the text and the ranges of the major herds of barren-ground caribou in the Northwest Territories.

(Ferguson and Gauthier 1992). The George River Herd of woodland caribou (*R. t. tarandus*) in eastern Canada was large in the 1880s, then declined sharply until the 1920s. The herd increased at an annual rate of 14% between 1954 and 1984, but it may have started to decline by the late 1980s (Messier et al. 1988).

Alaskan herds are also characterized by wide variations in herd size: The Forty Mile herds have varied between less than 10,000 to more than 50,000 this century (Valkenburg et al. 1994). It is difficult to determine causation retrospectively because several factors were acting simultaneously. The declines in the early 1950s and late 1960s coincided with unfavorable weather, but it is impossible to distinguish between the effects of winter and summer on the caribou population dynamics. Caribou populations on Arctic islands also fluctuate greatly. On Banks Island, numbers declined from 11,000 in 1972 to 1,000 in 1994 (Figure 6.6; Mclean and Fraser 1992; J. Nagy, 1995, personal communication). Inuit describe caribou as being more numerous in the 1960s and scarce earlier in the century. In the High Arctic, caribou also declined 90% between 1963 and 1974 with no evidence of subsequent recovery (Miller 1991).

The cause of caribou declines on the Arctic islands is conjectural, although we know that declines coincided with a significant increase in annual snowfall (Gunn et al. 1991). On Banks Island, early winter snowfall peaked in two years when many caribou died from malnutrition, presumably because foraging was difficult (Figure 6.7). The increase in snowfall has coincided with an earlier melt since the 1960s and warmer summers (Foster 1989; Government of NWT unpublished report).

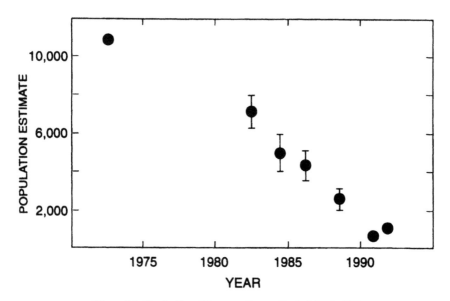

Figure 6.6. The decline of Peary caribou on Banks Island, NWT.

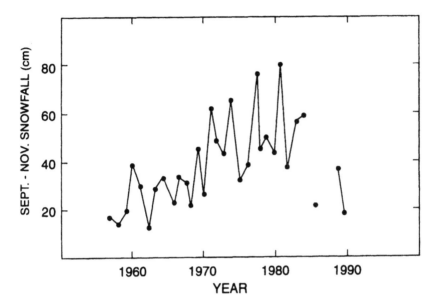

Figure 6.7. Early winter total snowfall at Sachs Harbor, Banks Island, NWT.

Periodic die-offs during catastrophic winters were believed to drive the population dynamics of caribou on Arctic islands (e.g., Gunn et al. 1981; Miller 1991). Population studies of reindeer (*R. t. platyrhyncus*) on the Arctic island of Svalbard (northern Europe) suggest that the situation may be more complex (Tyler 1987). High mortality (-16%, -17%, -47%) during three winters coincided with low summer rainfall preceding a cold winter and high reindeer densities (>2.7 reindeer/km²). When densities exceeded 5.1 reindeer/km² in the 1983–84 winter, mortality and emigration were greater, although the winter was less severe. Winter ranges on Svalbard are exposed heath and scree communities with shallow snow. Productivity of these dry communities is limited by summer moisture. Svalbard reindeer, despite large fat accumulations by autumn, rely on winter foraging for 75% of their energy requirements. The relative impacts of snow and ice conditions and summer forage growth on variations in reindeer winter diet are unclear (Tyler 1987).

Tyler's (1987) conclusions for reindeer serve as a cautionary note regarding the assumption that winter severity drives caribou dynamics on Arctic islands. Like the Svalbard reindeer, Peary caribou in winter feed on exposed plant communities dominated by dwarf cushion plants whose primary productivity is low and moisture-limited. The role of summer rainfall and caribou population densities in determining absolute forage availability, and the effect of winter conditions on relative availability, must be considered in determining the effect of severe winters on Peary caribou. Lichens are also characteristic of dry sites on continental tundra. These sites accumulate less snow, and lichens are easily accessible. In forests, caribou seek lichens during the winter, and their depletion is often associated with caribou malnutrition, especially during severe winters in maritime climates (Skogland 1985)

Severe winters may not act alone as a density-independent phenomenon limiting island populations of caribou without letting them reach high enough densities to reduce their forage. Peary caribou decreased on Banks Island during the 1970s when populations were high, summer precipitation was below average, and winter precipitation was above average. The combination of these three conditions could be detrimental, although there are insufficient data on interactions between caribou densities and weather effects on winter forage availability.

Weather was assumed to affect caribou populations on Arctic islands and in northern tundra or maritime climates. However, fluctuations in numbers of barren-ground caribou have generally been attributed to predators, hunting, and forage supplies (e.g., Bergerud 1983; Skogland 1991; Messier et al. 1988). Forage can be reduced by both high densities of caribou and winter weather. In Europe, new-born calves died when high densities of reindeer reduced forage availability for pregnant cows during winter (Skogland 1985). The effect of severe winters (specifically deeper snowpack in the 1950s) in compounding low forage availability (which coincided with high caribou densities) may also explain why the George River herd decreased (Crête and Payette 1990).

Fluctuations in musk-ox populations

We lack the historical perspective to gauge the timing and extent of population fluctuations on the continental mainland. At times, musk-oxen were numerous enough to be important in the aboriginal cultures on the mid-Arctic islands and mainland. Sharp declines on the NWT mainland followed unregulated commercial harvest in the late 1800s and early 1900s (Barr 1991), and numbers remained low until the 1970s, when populations started to increase.

Musk-oxen virtually disappeared from Banks and Victoria islands in the late 1800s. The decline preceded changes in hunting patterns (following the whalers and traders) by indigenous people but coincided with possible weather-related catastrophes (Gunn et al. 1991). Musk-oxen on Banks Island rebounded from less than 1,000 before 1970 to 3,000 in 1972 and then to 64,000 in 1994 (Gunn et al. 1991; J. Nagy, 1995, personal communication).

Most musk-ox ecology and population dynamics data are from their extreme northern range in the High Arctic Islands (Hubert 1977; Gray 1987). Hubert (1977) describes the dynamics as boom or bust, with reproductive failures and adult mortality associated with severe winters in the early 1950s, late 1960s, and early 1970s (Miller et al. 1977; Gray 1987). A similar situation may hold for Greenland: Musk-oxen died during mild winters with heavy snowfall and freezing rain (Forchhammer and Boertmann 1993). It was assumed, implicitly or explicitly, that periodic catastrophic winters limited musk-ox numbers.

Single events, such as exceptionally high snowfall, correlate with reduced calf production and survival. Older musk-oxen also die when ice and snow restrict forage availability (Miller et al. 1977; Gray 1987; Forchhammer and Boertmann 1993); bulls and subadults often have the greatest losses (Gunn et al. 1989). Miller et al. (1977) estimated that numbers of musk-oxen on one Arctic island declined by 70% between late March and late August 1974, with the decline attributed to deaths following a severe winter.

Caribou and musk-ox responses to environmental variation

Fluctuations in ungulate numbers are usually attributed to predation or food limitation, with weather relegated to special cases. In North America, the notion that food rarely limits caribou is entrenched, but that conclusion stems from coincidences between population trends and predator numbers or some presumed cause.

An alternative approach focuses on data from an herbivore-forage-weather system for kangaroos (*Macropus giganteus*) in semi-arid grasslands of Australia. Highly variable and unpredictable annual rainfall determines forage biomass, which in turn drives the rates of change in the kangaroo population. Kangaroos are not completely at the mercy of the climate. Higher kangaroo densities reduce forage biomass, and this feedback imposes a weak but effective brake on the system (Caughley 1987). The term *centripetality* conveys how feedback loops (herbivore-forage biomass, forage biomass-forage growth) push the system toward equilibrium to counteract fluctuations in weather. Short-term variation in weather and plant biomass explained longer-term (decades) fluctuations in kangaroo numbers.

There are similarities between the system dynamics of caribou and kangaroos. The Arctic climate is unpredictably variable. On Banks Island, the coefficient of variation for snowfall at the end of May is 87% and for the length of the growing season is 47%; serial correlations between years are nonsignificant (Caughley and Gunn 1993). Weather and its effect on forage could drive fluctuations in numbers of caribou over decades. Wolves and other predators can accentuate or dampen the fluctuations.

A prominent feature of Arctic ecosystems is that annual productivity is pulsed and most nutrients are held in dead plant material (Chapter 5). Nutrient input in tundra ecosystems is low, and nutrient cycling through soil organic matter is slow. Plants are adapted by internal recycling of nutrients from leaves into belowground (storage) tissue. Herbivory plays a role in nutrient cycling through the rapid breakdown of herbivore dung, which makes nutrients quickly available for plant uptake. Grazing initially has a positive feedback as the released nutrients stimulate productivity (Jefferies 1992). As herbivore densities increase, the removal of forage increases beyond the capability of plants to compensate.

Caribou populations are characterized by fluctuations in their numbers over decades. Surges of recovery and decline are predictable and understandable in terms of a variable climate, its effects on forage supplies, and the effects of forage on the rate of herbivore increase. Effects of weather are not always immediate or conspicuous; relatively minor weather events may have prolonged and cascading population effects. Fluctuations are more extreme where climate restrictions on forage are tighter. Arctic islands such as the High Arctic Islands and northern Greenland approach the limits of herbivore habitat in terms of length of plant growing season and productivity. At the extreme of potential ungulate habitat, weather fluctuations are likely to have a greater impact on population dynamics.

Climate change, people, caribou, and musk-oxen

Uncertainties in future climate and in the population ecology of ungulates make it difficult to determine how caribou and musk-ox numbers might change, and how these changes would affect people. Predicting changes in absolute forage availability is complex, given the diversity of plant species and communities. Warmer temperatures, a longer growing season, and more rapid nutrient cycling may increase net productivity, although species response would differ (Chapter 5). Photoperiod constraints, especially in autumn, may limit plant response to a longer growing season. Warmer and moister summers may lead to more storms with lightning strikes that ignite forest fires. Although forest fires are part of the ecology of barren-ground caribou, an accelerated burn rate could lead to loss of winter forage. Increased evapotranspiration and a lower water table may decrease plant productivity on drier sites. Those sites are also less likely to see an accentuated nutrient cycle; lower biomass reduces the amount of time animals spend at the site and reduces fecal accumulation. Drier sites usually have less snow accumulation and are preferred winter feeding sites for caribou.

A longer, warmer summer will affect caribou directly, as well as affect their food supply. Delays in freeze-up will change migration patterns as some caribou herds are forced to detour around large lakes. More deaths are possible from predation and from caribou breaking through ice. Increases in insect harassment will markedly reduce caribou body condition in summer.

In winter, warmer temperatures and increased snowfall would decrease forage availability. Crusting, ice layers, and denser snow from freeze-thaw cycles are more costly in terms of energy use when compounded with increased snow. Earlier snowmelt would compensate for some reduction in relative availability of forage caused by snow. An earlier springtime melt could reduce the energy cost of spring migration if it occurred before migration. However, if greater winter snowfall prolongs the melt, spring migration would be an energetic drain if caribou have to trudge through deep, wet, or crusted snow.

Although the amount of summer forage through plant growth (absolute availability) may increase, the amount available in winter as mediated by snow (relative availability) will become energetically more costly. Lower winter forage availability and insect harassment will almost certainly increase energy costs, with possible decreases in fecundity (fertility and births) and greater mortality. Increased snow crusting would favor more wolf predation. Individual changes may be small, but their cumulative effects would have a significant impact on survival and fecundity. Climate change will have greater effects in areas where the system is near its limits. Those limits may be where herbivore densities are high relative to forage or climate is already limiting (e.g., on Arctic islands). Climate change will cause the greatest perturbation where the margin between survival and mortality is the slimmest.

It is unclear how climate affects the relationship between caribou and musk-oxen. Many Inuit believe that caribou decline as musk-oxen increase, but biologists are less sure. Musk-ox and caribou populations have different population trajectories in different areas. Caribou and musk-oxen both increased on the NWT mainland and Victoria Island during the 1970s (Gunn 1990; Barr 1991; Ferguson and Gauthier 1992). However, the sharp decline in caribou on Banks Island since the 1970s coincided with an increase in musk-oxen.

The different trends in population sizes of the two species suggest that their trajectories are independent, which is not surprising given the differences in their ecology and anatomy. Musk-oxen are large-bodied grazers able to forage selectively to maximize their intake of forage in the plant growing season. The large rumen and omasum (first and third chambers of an ungulate's digestive system) (J. Adamczewski, 1993, personal communication) allow musk-oxen to digest poor quality winter forage. Musk-oxen can therefore survive all but the most severe winters.

Caribou and musk-oxen will likely be more influenced by their independent responses to weather than by interaction between the two species. Warmer, moister summers will benefit sedge meadows where musk-oxen (but not caribou) forage. Warmer, moister winters would be detrimental for musk-oxen and could temporarily change the relationship between the two species. Deeper snow may force the musk-oxen to crater for forage on the same slopes where caribou forage; this would cause packed, hard snow that would handicap caribou. A similar change in competitive interactions occurs among moose (*Alces alces*), elk (*Cervus elaphus*), and white-tailed deer (*Odocoileus virginianus*) during winters with heavy snow (Jenkins and Wright 1987).

Inuit have coped with climate change throughout their history in North America. Now that most Inuit live in permanent communities, migrating to another location or hunting different species are not viable options. More caribou are taken during winter than in any other season, and shorter and warmer winters with more snow may impede hunting. Any amelioration of the climate will probably foster human activity in the Arctic. Increased shipping, more all-weather roads, and more aircraft could influence ungulate abundance and distribution. Interruption of inter-island animal movement is an obvious potential impact, but more subtle responses, such as higher energy costs from responding to disturbance, will be difficult to assess.

Summary

Despite uncertainty about future ecological relationships, the most likely outcome of a warmer climate will be a decline in caribou and musk-oxen, particularly if there is a greater frequency and amplitude of weather extremes. This prediction can be made with greater confidence for ungulates on Arctic islands and in regions with continental maritime climates. This chapter has emphasized the dynamics of ungulates and forage, but ecological relationships with parasites and predators will also be affected by a warmer climate. In the NWT, a growing human population and changing economic expectations are increasing wildlife use. If a warmer climate is unfavorable to caribou and musk-oxen, their populations may be unable to support higher levels of harvesting, leading to problems for both the animals and the people who depend on them.

References

Barr, W. 1991. *Back from the brink: The road to muskox conservation in the Northwest Territories*. Komatik Series 3. Calgary: Arctic Institute of North America, University of Calgary.

Bergerud, A. T. 1983. The natural population control of caribou. In *Proceedings of the symposium on natural regulation of wildlife populations*, edited by F. L. Bunnell, D. S. Eastman, and J. M. Peek, 14–61. Moscow, ID: Forest Wildlife and Range Experiment Station, University of Idaho.

Bureau of Statistics. 1994. Government of the Northwest Territories, Yellowknife, NWT. *Statistics Quarterly* 16:1–62.

Callaghan, T. V., B. Å. Carlsson, and N. J. Tyler. 1989. Historical records of climate-related growth in *Cassiope tetragona* from the Arctic. *Journal of Animal Ecology* 77:823–837.

Caughley, G. 1987. Ecological relationships. In *Kangaroos: Their ecology and management in the sheep rangelands of Australia*, edited by G. Caughley, N. Shepherd, and J. Short, 159–187. Cambridge: Cambridge University Press.

Caughley, G., and A. Gunn. 1993. Dynamics of large herbivores in deserts: Kangaroos and caribou. *Oikos* 67:47–55.

Crête, M., and S. Payette. 1990. Climate changes and caribou abundance in northern Quebec over the last century. *Rangifer* 3:159–165.

Eloranta, E., and M. Nieminen. 1986. Calving of the experimental reindeer herd in Kaamanen during 1970–1985. *Rangifer* Special Issue 1:115–122.

Fancy, S., and R. G. White. 1985. Energy expenditure by caribou while cratering in the snow. *Journal of Wildlife Management* 49:987–993.

Ferguson, M. A. D., and L. Gauthier. 1992. Status and trends of *Rangifer tarandus* and *Ovibos moschatus* populations in Canada. *Rangifer* 12:127–141.

Forchhammer, M., and D. Boertmann. 1993. The muskoxen *Ovibos moschatus* in north and northeast Greenland: Population trends and the influence of abiotic parameters on population dynamics. *Ecogeography* 16:299–308.

Foster, J. L. 1989. The significance of the date of snow disappearance on the arctic tundra as a possible indicator of climate change. *Arctic and Alpine Research* 21:60–70.

Gray, D. R. 1987. *The Muskoxen of Polar Bear Pass.* Markham, Ontario: Fitzhenry and Whiteside.

Gunn, A. 1990. The decline and recovery of caribou and muskoxen on Victoria Island. In *Canada's missing dimension: Science and history in the Canadian Arctic Islands*, edited by C. R. Harrington, 590–607. Ottawa: Canada Museum of Nature.

Gunn, A., B. Mclean, and F. L. Miller. 1989. Evidence for and possible causes of increased mortality of adult male muskoxen during severe winters. *Canadian Journal of Zoology* 67:1106–1111.

Gunn, A., C. C. Shank, and B. McLean. 1991. The history, status and management of muskoxen on Banks Island. *Arctic* 44:188–195.

Gunn, A., F. L. Miller, and D. C. Thomas. 1981. The current status and future of Peary caribou (*Rangifer tarandus pearyi*) on the Arctic Islands of Canada. *Biological Conservation* 19:283–296.

Helle, T., and L. Tarvainen. 1984. Effects of insect harassment on weight gain and survival in reindeer calves. *Rangifer* 4:28–34.

Hubert, B. A. 1977. Estimated productivity of muskox in Truelove Lowland. In *Truelove Lowland, Devon Island, Canada—A High Arctic ecosytem*, edited by L. C. Bliss, 467–491. Edmonton: University of Alberta Press.

Jefferies, R. L. 1992. Tundra grazing systems and climatic change. In *Arctic ecosystems in a changing climate: An ecophysiological perspective*, edited by F. S. Chapin, R. L. Jefferies, J. F. Reynolds, G. R. Shaver, J. Svoboda, and E. W. Chu, 391–412. New York: Academic Press.

Jenkins, K. J., and R. G. Wright. 1987. Dietary niche relationships among cervids relative to snowpack in northwestern Montana. *Canadian Journal of Zoology* 65:1397–1401.

Maxwell, B. 1980. *The climate of the Canadian Arctic Islands and adjacent waters.* Ottawa: Atmospheric Environment Service, Environment Canada.

Maxwell, B. 1992. Arctic climate: Potential for change under global warming. In *Arctic ecosystems in a changing climate: An ecophysiological perspective*, edited by F. S. Chapin, R. L. Jefferies, J. F. Reynolds, G. R. Shaver, J. Svoboda, and E. W. Chu, 11–34. New York: Academic Press.

McGhee, R. 1990. The peopling of the Arctic Islands. In *Canada's missing dimension: Science and history in the Canadian Arctic Islands*, edited by C. R. Harrington, 666–676. Ottawa: Canada Museum of Nature.

Mclean, B. D., and P. Fraser. 1992. *Abundance and distribution of Peary caribou and muskoxen on Banks Island, NWT, June 1989.* NWT Wildlife Service File

Report 106. Yellowknife: Department of Renewable Resources, Government of Northwest Territories.

Messier, F., J. Huot, D. le Henaff, and S. Luttich. 1988. Demography of the George River Herd: evidence of population regulation by forage exploitation and range expansion. *Arctic* 41:279–287.

Miller, F. L. 1991. *Peary caribou—A status report*. Ottawa: Canadian Wildlife Service.

Miller, F. L., and A. Gunn. 1986. Observations of barren-ground caribou travelling on thin ice during fall migration. *Arctic* 39:85–88.

Miller, F. L., R. H. Russell, and A. Gunn. 1977. *Distributions, movements and numbers of Peary caribou and muskoxen on Western Queen Elizabeth Islands, Northwest Territories, 1972–74*. Report Series 40. Ottawa: Canadian Wildlife Service.

Muc, M. 1977. Ecology and primary production of sedge-moss communities, Truelove Lowland. In *Truelove Lowland, Devon Island, Canada—A High Arctic ecosytem*, edited by L. C. Bliss, 157–184. Edmonton: University of Alberta Press.

Parker, K., R. G. White, M. P. Gillingham, and D. F. Holleman. 1990. Comparison of energy metabolism in relation to daily activity and milk consumption by caribou and muskox neonates. *Canadian Journal of Zoology* 68:104–114.

Russell, D. R., A. M. Martell, and W. Nixon. 1993. Range ecology of the Porcupine Caribou Herd in Canada. *Rangifer* Special Issue 8:168.

Skogland, T. 1985. The effects of density-dependent resource limitation on the demography of wild reindeer. *Journal of Animal Ecology* 54:359–374.

Skogland, T. 1991. What are the effects of predators on large ungulate populations? *Oikos* 61:401–411.

Svoboda, J. 1977. Ecology and primary production of raised beach communities, Truelove Lowland. In *Truelove Lowland, Devon Island, Canada—A High Arctic ecosystem*, edited by L. C. Bliss, 185–216. Edmonton: University of Alberta Press.

Thomas, D. C. 1982. The relationship between fertility and fat reserves of Peary caribou. *Canadian Journal of Zoology* 60:597–602.

Tyler, N. J. C. 1987. Body composition and energy balance of pregnant and non-pregnant Svalbard reindeer during winter. *Symposium Zoological Society of London* 57:203–229.

Valkenburg, P., D. G. Kellyhouse, J. L. Davis, and J. M. Ver Hoef. 1994. Case-history of the Fortymile Caribou Herd, 1920–1990. *Rangifer* 14:11–22.

White, R. G., D. F. Holleman, and B. A. Tiplady. 1989. Seasonal body weight, body condition and lactational trends in muskoxen. *Canadian Journal of Zoology* 67:1125–1133.

White, R. G., F. L. Bunnell, E. Gaare, T. Skogland, and B. Hubert. 1981. Ungulates on arctic ranges. In *Tundra ecosystems: A comparative analysis*, edited by L. C. Bliss, W. Heal, and J. J. Moore, 397–483. Cambridge: Cambridge University Press.

7

Effects of climate change on marine mammals in the Far North

Kathryn A. Ono

Predicting the effect of climate change on marine mammals is a monumental task, similar to predicting if climate change will affect the marine environment at all. Marine mammals living in the North Pacific include members of three orders and 13 families (the Steller sea cow [*Hydrodamalis gigas*], a member of the order Sirenia, was extirpated by 1758 A.D.). Some species are sedentary and local in distribution, while others migrate from warm, temperate breeding grounds thousands of kilometers away to summer feeding grounds among the Arctic ice. Marine mammals feed on zooplankton, larger invertebrates, fish, and other marine mammals.

The only available evidence of how a warmer climate might affect marine mammals is from observations of populations during short-term temperature fluctuations. The effect of climatic variation has been documented for only a few species. This chapter provides an overview of the marine mammals inhabiting the North Pacific and then reviews the effects of environmental temperature change on marine mammals in terms of physiological tolerances and effects on prey abundance. Potential effects due to habitat loss are discussed for the various species. Using short-term fluctuations as a model, general predictions are offered on how long-term oceanic warming may impact marine mammals in the North Pacific.

Marine mammals of the North Pacific

Mammals are warm-blooded vertebrates that produce milk for their neonates (newborns) in specialized structures called mammary glands and have hair, except in cases in which it is secondarily lost as in the cetaceans (which include whales, dolphins, and porpoises). Marine mammals have special adaptations for marine conditions and have an obligate link to the marine environment (Estes 1991).

Table 7.1. Marine mammals occurring in the far North Pacific.

Order/Family	Common name	Scientific name	Migratory	Distribution in North Pacific	Diet	Human interactions
Cetacea (Suborder: Mysticeti) Balaenidae	Bowhead whale	*Balaena mysticetus*	Yes	(S) Arctic ice edge, (W) Bering Sea	Zooplankton	Were commercially exploited
	Northern right whale	*Eubalaena glacialis*	Vague	Bering Sea to Baja	Zooplankton	Were commercially exploited
Balaenopteridae	Blue whale	*Balaenoptera musculus*	Yes	(S) Gulf of AK to E. Aleutians, (W) Baja, Gulf of CA	Krill	Were commercially exploited
	Fin whale	*Balaenoptera physalus*	Yes	(S) Bering Sea, (W) Baja	Krill, fish	Were commercially exploited
	Sei whale	*Balaenoptera borealis*	Yes	Gulf of AK to Channel Islands	Plankton to small fish	Were commercially exploited
	Minke whale	*Balaenoptera acutorostrata*	Yes	(S) Chukchi Sea to Baja, (W) central CA to equator	Fish, krill, copepods	Historical harvest Japan, Korea; some harvest by AK Natives
	Humpback whale	*Megaptera novaeangliae*	Yes	(S) Bering Strait, (W) Baja, Hawaii	Krill, schooling fish	Were commercially exploited
Eschrichtiidae	Gray whale	*Eschrichtius robustus*	Yes	(S), (F) Bering, Chukchi, Beaufort Seas, CA (S)	Benthic amphipods and crustaceans	Were commercially exploited; Native subsistence harvest
(Suborder: Odontoceti) Physeteridae	Sperm whale	*Physeter macrocephalus*	Yes	(S) Bering Sea, (W) central CA	Squid, fish	Were commercially exploited
Monodontidae	Beluga (white whale)	*Delphinapterus leucus*	Some populations	(S) Beaufort Sea, (W) Bering Sea	Crustaceans to schooling fish	Were commercially exploited

Ziphiidae	Baird's beaked whale	*Berardius bairdii*	Possibly further south in winter	N. California to Pribilof Islands	Deep-sea fish, octopus, squid	Small historic and present commercial harvest
	Cuvier's beaked whale	*Ziphius cavirostris*	Uncertain	So. Bering Sea to New Zealand	Squid, deep-water fish	Some incidental mortality in fisheries
	Stejneger's beaked whale	*Mesoplodon stejnegeri*	Unknown	So. Bering Sea, south to Monterey, CA	Squid, fish	Few
	Hubbs' beaked whale	*Mesoplodon carlhubbsi*	Unknown	Vancouver Island to San Diego, CA	Squid, fish	Few
Delphinidae	False killer whale	*Pseudorca crassidens*	Uncertain	Prince William Sound to so. CA	Squid, large fish	Taken for food in Japan; incidental takes in tuna fishery
	Killer whale	*Orcinus orca*	Resident and transient populations	Worldwide in coastal waters	Marine mammals, birds, turtles, fish, squid	Small commercial fishery; public display; competition with other fisheries; high contaminant levels
	Short-finned pilot whale	*Globicephala macrorhynchus*	Offshore to inshore in spring	Mostly tropical, but seen as far north as Gulf of AK	Squid	Small harvest in tropics
	Pacific white-sided dolphin	*Lagenorhynchus obliquidens*	Possibly inshore/offshore	Kamchatka Peninsula, Kodiak Island, AK south to Baja	Fish, squid	Some for food in Japan; public display; incidental takes in drift and gill nets

(continued)

Table 7.1. *Marine mammals occurring in the far North Pacific. (Continued)*

Order/Family	Common name	Scientific name	Migratory	Distribution in North Pacific	Diet	Human interactions
	Risso's dolphin	*Grampus griseus*	Unknown	Offshore SE Alaska and Kuriles to central Chile	Squid, fish	Small incidental take in fisheries
	Northern right whale dolphin	*Lissodelphis borealis*	(W), (Sp) south and inshore; (S), (F) north and offshore	British Columbia to Baja, mostly deep water	Squid, fish	Fishery in Japan; incidental in gill-net and seine fish-eries; some on public display
Phocoenidae	Harbor porpoise	*Phocoena phocoena*	Inshore/ offshore	Coastal Beaufort Sea to Pt. Con-ception, CA	Fish, cephalopods	Some subsistence harvest; many in incidental entan-glement; high con-taminant levels
	Dall's porpoise	*Phocoenoides dalli*	Yes	South Bering Sea, Pribilof Islands to Baja	Squid, crus-taceans, fish	Historical large Japanese fishery; incidental entan-glement
Pinnipedia Phocidae	Northern ele-phant seal	*Mirounga angustirostris*	Yes (after breeding and again after molt)	Aleutian Islands, Gulf of AK south to Baja	Fish, cephalopods	Historical commer-cial fishery left few survivors; incidental offshore fishery takes
	Bearded seal	*Erignathus barbatus*	Yes (within Arctic/ subarctic)	(S) in Chukchi, Beaufort Seas, (W)/(Sp) in Bering Sea (follow ice)	Bivalves, crustaceans, cephalopods; fish	Large Russian com-mercial fishery; AK Native subsis-tence harvest

Common name	Scientific name	Migratory	Distribution	Diet	Threats/uses
Ribbon seal	*Phoca fasciata*	Yes (limited)	Chukchi to Bering Sea (follow ice)	Crustaceans, fish, krill, cephalopods	Russian commercial fishery; AK Native subsistence takes
Largha (spotted) seal	*Phoca largha*	Yes (limited)	Chukchi to Bering Sea (follow ice)	Fish, cephalopods, crustaceans	Large Native subsistence harvest; competition with commercial fisheries
Ringed seal	*Phoca hispida*	Yes (limited)	Bristol Bay to Chukchi, Beaufort Seas (follow ice)	Fish, crustaceans, krill	Large Native subsistence take
Harbor seal	*Phoca vitulina richardsi*	Local movements	Pribilof Islands to so. CA	Fish, cephalopods, krill	Fishery conflicts/entanglement; accumulated pollutants; probably some AK Native subsistence use
Otariidae					
California sea lion	*Zalophus californianus*	Yes, especially males	SE AK to Baja	Fish, cephalopods	Historical commercial exploitation; conflicts with commercial and sport fisheries
Steller sea lion	*Eumetopias jubatus*	Seasonal movement, probably some migration	Bering Sea to southern CA	Fish, cephalopods, crustaceans, pinnipeds	Small subsistence harvest by AK Natives; incidental fisheries takes; competition with commercial fisheries
Northern fur seal	*Callorhinus ursinus*	Yes	Commander, Pribilof Islands to southern CA	Fish, cephalopods	Historical commercial exploitation

(continued)

109

Table 7.1. Marine mammals occurring in the far North Pacific. (Continued)

Order/Family	Common name	Scientific name	Migratory	Distribution in North Pacific	Diet	Human interactions
Obodenidae	Walrus	*Odobenus rosmarus divergens*	Yes (limited)	Chukchi Sea to Bristol Bay	Bivalves, fish, cephalopods, misc. invertebrates	Historical commercial harvest; Russian commercial harvest; AK Native subsistence takes
Carnivora Ursidae	Polar bear	*Ursus maritimus*	Yes (limited)	Bering Sea to Chukchi, Beaufort Seas (follow ice)	Pinnipeds, other mammals, vegetation, carrion	Historical sport hunting; subsistence harvest by AK Natives; public display
Mustelidae	Sea otter	*Enhydra lutris*	No	Coastal Aleutian Islands to southern CA	Various invertebrates	Heavy historical commercial exploitation; subsistence harvest by AK Natives; some entanglement

Sources: Leatherwood and Reeves (1983), Riedman (1990), and Wynne and Folkens (1993).

Notes: Notations for distributions and breeding habitat: (S) = summer, (F) = fall, (W) = winter, (Sp) = spring. Baja refers to Baja California, Mexico. CA denotes California; AK denotes Alaska.

Marine mammals inhabiting the North Pacific are members of three different orders: the Cetacea, Pinnipedia, and Carnivora (Table 7.1; see this table for scientific names of marine mammals discussed in this chapter). Cetaceans are completely adapted for aquatic life and carry out all phases of their lives in water (e.g., Evans 1987). They have a streamlined, fusiform body shape and no hind limbs. Their tail is modified into a flattened fluke used for propulsion. Except for a few whiskers on some species, they have minimal hair in order to reduce drag while moving underwater. Nostrils in cetaceans are in the tops of their heads to aid in air exchange during brief visits to the surface. Cetaceans rely on a subcutaneous layer of blubber (fat) to insulate them from the high thermal conductivity of cold North Pacific water (water is 25 times more conductive than air).

Members of the suborder Mysticeti have a specialized feeding structure called *baleen* with which they filter out prey items from large quantities of water engulfed in their mouths. This suborder includes the largest animals on Earth, including the blue whale weighing over 100 metric tons; the smallest species is the minke whale weighing 8 metric tons. The other cetacean suborder is the Odontoceti, or toothed whales. This suborder includes the killer whale, which at times preys on pinnipeds and larger cetaceans; it also includes the harbor porpoise, a fish-eating cetacean weighing only 60 kilograms. Twenty-two cetacean species inhabit the North Pacific (Table 7.1).

The pinnipeds are amphibious; all feed in the ocean but they must return to land to give birth to their pups, and some rely on ice for parturition (giving birth) and pup rearing. As a result, they can be affected by climate in both terrestrial and marine habitats. During the breeding season, females return to traditional pupping and breeding areas. All species have a postpartum estrus (a period of sexual receptivity after giving birth), occurring a few days to several weeks after the birth of a pup. Males congregate on land or in the water near shore to inseminate estrous (being in an estrus condition) females. To move on both land and in water, pinnipeds have retained both front and hind limbs. The hind limbs of phocids (true seals) are webbed for propulsion in water. Otariids (eared seals—the sea lions and fur seals) use large, flat forelimbs for aquatic locomotion (Figure 7.1). Pinnipeds have retained their hair, which offers some insulation on land. The northern fur seal has thick fur that offers insulation on both land and in water. The remainder of the far North Pacific pinnipeds use subcutaneous blubber as their primary insulation against cold water (Figure 7.2). The same insulation that keeps pinnipeds warm in water can cause overheating on land if temperatures are too high; this would be analogous to a human basking in the sun in a thick wetsuit. The diet of North Pacific pinnipeds ranges from zooplankton to (occasionally) other pinnipeds. Eleven pinniped species inhabit the North Pacific (Table 7.1).

Sea otters, like cetaceans, carry out all life functions in the water, including parturition and caring for young. Sea otters rely on a thick fur coat for insulation against cold water and constantly groom themselves to keep a good air layer trapped within the heavy fur. To keep warm, sea otters must maintain a high meta-

Figure 7.1. Harbor seals hauled out on a rocky beach, Año Nuevo Island, California. Photo by Kathryn Ono.

bolic rate, requiring about 20–25% of their body weight in energy intake per day. They are a coastal species, largely remaining over shallow water where their invertebrate prey live. Sea otters are the smallest marine mammal, with adults weighing only 35 kilograms (VanBlaricom and Estes 1988; Estes 1991).

Polar bears (*Ursus maritimus*) are the most recently evolved marine mammal, being primarily adapted to life on the ice. Their principal prey are seals, which they catch on the ice (not in the water) or at the seals' breathing holes (Stirling 1988).

Direct effects of increased temperatures

Studies relating to the effects of increased temperatures concern two topics: (1) range expansion of warm water species and range contraction of cold water species, and (2) impacts of reduced habitat because of changes in the distribution and character of ice and island beaches.

There are two documented accounts of cetacean species expanding their range during periods of warmer than usual water temperatures. Wells et al. (1990) observed a northward range expansion of bottlenose dolphins of more than 600 kilometers during the 1982–83 El Niño event. Dolphins were seen in this northern range until at least 1988. Although the authors were not certain the dolphins were responding to warmer waters or merely following prey, the ability of many species to expand their ranges suggests that expansion may occur as a result of increased ocean temperatures. Leatherwood et al. (1980) found that movements of Risso's dolphins, at higher latitudes and in deeper water off the continental shelf at lower latitudes, were related to warmer surface water temperatures. Seasonal movements on and offshore as well as long-term changes in range have occurred in this species in response to environmental fluctuations. Pacific white-sided dolphins and the right whale dolphin also follow warm water seasonally and remain in more northerly waters throughout unseasonably warm years (Leatherwood and Reeves 1978; Leatherwood and Walker 1979).

Most of the cetacean species commonly found in the far North Pacific are wide-ranging and/or migratory (Table 7.1). The large range of temperatures in which most are found suggests they could thermoregulate normally in a warmer North Pacific. At most risk from thermal stress are those species that remain at high latitudes throughout the year, such as the bowhead whale (already severely depleted by humans) and beluga.

Direct effects of warmer conditions are poorly documented for pinnipeds. Pinnipeds must haul out onto land for parturition and pup rearing, and it is difficult for them to dissipate heat on land due to lower thermal conductivity of air compared to water. In addition, pinnipeds do not have perspiratory glands, and most species do not pant (Riedman 1990). They have a well-developed network of capillaries in the skin and flippers through which blood transfers heat from the

Figure 7.2. Steller sea lions on a breeding site, Lowrie Island, Alaska. Photo by Kathryn Ono.

warm core of the animal into the air. Pinnipeds wave flippers in the air (mostly otariids) or dig into the cool sand and flip wet sand on their backs when hot (phocids). Some species (e.g., Guadalupe fur seals [*Arctocephalus townsendi*]) search out caves and boulders for shade. When these methods do not cool animals sufficiently, they move into the water. Prior to estrus, females of some species (e.g., northern fur seals) are restricted by the male from leaving his territory. Very young otariid pups generally do not enter the water (personal observation). In these cases, increased air and substrate temperatures could cause overheating. On Marmot Island, Alaska, the substrate temperature of the dark cobble beaches, which are the Steller sea lion breeding habitat, can reach up to 53°C.

Changes in pinniped distribution have been correlated with periods of warm sea surface temperatures associated with El Niño conditions that cause a 3°–12°C increase in ocean temperature. An El Niño is an irregularly occurring climatological event in the Pacific Ocean, defined by above-normal sea temperatures off the coast of Peru and Ecuador. Weak El Niños cause a rise in sea surface temperature of less than 2°C at the equator, while very strong El Niños are associated with 7°–12°C increases. Strong events occur every 10 years on average (Quinn et al. 1987). The effects of strong to very strong events can extend to higher latitudes in the North Pacific; the very strong 1982–83 El Niño may have affected ocean temperatures as far north as the Bering Sea, causing up to a 5°C increase in temperatures in the Gulf of Alaska (Royer 1985).

The El Niño is also associated with increases in sea level, deepening of the thermocline (a layer separating warmer water above from cooler water below), increased salinity, and decreased upwelling, leading to a decline in primary productivity and marine mammal prey. In areas influenced by the 1982–83 El Niño, many prey species migrated farther offshore, northward, or to greater depths to find cooler water. Although we are not certain that these changes were in response to a northward shift in prey distribution or to warmer ocean temperatures, unusual northward movements (into cooler water) of the northern fur seal, South American fur seal, and California sea lion were observed during and after the 1982–83 El Niño (DeLong and Antonelis 1991; Guerra and Portflitt 1991; Huber 1991).

All three otariid species are wide-ranging (Table 7.1). Increased temperatures during the breeding season might encourage establishment of more northerly breeding sites by shifting populations from the southern border of the distributions. An increased number of California sea lion pups were born north of established breeding colonies on Año Nuevo Island, California, during years affected by recent El Niños (Ono et al. in review). The decline in Steller sea lions, approximately 50% over the last 30 years (Loughlin et al. 1992), might also be partially due to an overall warming trend. However, not all populations have declined at the same rate, and some have actually increased over this time. Populations in the extreme southern portions of the range have the highest rates of decline. Steller sea lions are no longer present on the Channel Islands (southern California) and have declined more than 80% on Año Nuevo Island and the Far-

tral California) (LeBoeuf et al. 1991). Long-term warming trends, in addition to more frequent El Niño events over the last 30 years (Folland et al. 1990), may have exacerbated the decline in the southern parts of the species range. A comparison of Steller sea lions on Año Nuévo Island to a stable population in southeast Alaska shows a significantly higher frequency of heat dissipation behaviors in the more southerly population during the 1993 breeding season (Ono, unpublished data). Because the majority of this species has always occurred in higher latitudes, we might speculate that they are better adapted for cooler climates. If this were true, we might expect to see a decline in northern fur seals on the Channel Islands. However, the northern fur seal population on San Miguel Island (the only breeding population south of the Pribilof Islands, Alaska) has increased over the last 30 years (DeLong and Antonelis 1991).

Land-breeding pinnipeds (in contrast to those which breed on ice) use traditional breeding sites, primarily located on offshore islands, away from terrestrial predators. Many of these sites are on beaches and rocky outcroppings surrounded by high cliffs. Rising sea levels could decrease the size of breeding areas, as well as increase susceptibility to wave-washing during high tides. Increased wave-wash, especially during storms, could increase pup mortality. Northern elephant seal pup mortality increased on one crowded breeding area during intense winter storms in 1983 (Le Boeuf and Reiter 1991). Huge waves resulting from a hurricane during the 1992 Guadalupe fur seal breeding season caused large numbers of young pups to be washed out to sea. Most were able to get back on to land, but many died from starvation because their mothers were unable to find them (personal observation). Most of the phocids inhabiting the far North Pacific, as well as polar bears and walruses, are tied to the ice year round (Table 7.1). Changes in the abundance and distribution of ice could severely affect species that use ice for breeding and mating or make seals more vulnerable to predation by polar bears. Sea otters might be expected to extend their ranges northward with retreating ice (Fittner 1993).

Indirect effects due to changes in prey distribution

Although direct effects of climate change may not seriously impact many species of marine mammals in the Far North, changes in the abundance and distribution of prey would affect them all. Fittner (1993) examines the evidence for increased upwelling in coastal zones and possibly decreased oceanic upwelling under a long-term warming. This is good news for marine mammals that forage near the coast (e.g., sea otters, walrus, belugas, and harbor seals), if increased upwelling translates to increased productivity. Decreased oceanic upwelling would mean a decline in prey for marine mammals that feed in deep water, including most of the cetaceans. Top carnivores such as the polar bear may also suffer a decline in pinniped prey.

Some species are probably more capable of switching their diet under altered conditions than others. Prey switching in California sea lions (DeLong et al. 1991) and northern fur seals (DeLong and Antonelis 1991) occurred on the Channel Islands during the 1982–83 El Niño (Figure 7.3). For instance, northern anchovy comprised 5% of the diet of California sea lions in 1982 (unaffected by the El Niño) and 65% of the diet during the height of El Niño in 1983 (DeLong et al. 1991). Northern anchovy (*Engraulis mordax*) comprised 2% of the diet of northern fur seals in 1982 and over 25% of the diet in 1983 (DeLong and Antonelis 1991). Both populations suffered increased mortality during 1983, suggesting that although they were capable of changing their diet, different prey items are not necessarily interchangeable with respect to nutritional value.

Prey distribution and pinniped abundance in the eastern Pacific changed as a result of the 1982–83 El Niño (Trillmich et al. 1991). The magnitude of the effects on pinniped populations was positively correlated with the magnitude of the perturbation, with populations near the equator (center of El Niño) suffering more than those at higher latitudes. On the Channel Islands, northern fur seals suffered increased mortality of pup and juvenile age classes. The number of adult males and females returning to breed decreased by 20% and 50%, respectively, and the number of pups born during the El Niño declined 60%. Pups grew at a slower rate, and the duration of female feeding trips increased. The effects of the El Niño on California sea lions, also breeding on the Channel Islands, were similar (Ono et al. 1987; Francis and Heath 1991), and female sea lions expended a

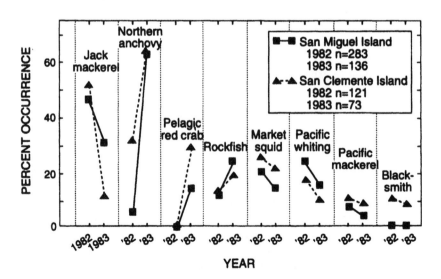

Figure 7.3. Summer diet of California sea lions on San Miguel and San Clemente Islands, California, 1982–83. Adapted from Delong et al. (1991), figure 2. Reprinted with permission of Springer-Verlag.

greater amount of energy during foraging trips in the El Niño year compared to normal years (Costa et al. 1991). More time and energy spent in prey capture meant less milk produced for pups, which was reflected in their slower growth rates and higher mortality due to starvation.

Phocid seals, such as the northern elephant seal, were also affected. In addition to increased mortality, juvenile mortality increased for those born the year prior to, the year of, and the year after the El Niño (Huber et al. 1991; LeBoeuf and Reiter 1991). This mortality and the increased time females spent at sea feeding after lactation and prior to molting (LeBoeuf and Reiter 1991; Stewart and Yochem 1991) indicate that elephant seals were food-stressed during this period, especially the younger, less experienced foragers. Northern elephant seals feed off the continental shelf south of the Aleutian Islands and Gulf of Alaska (Stewart and DeLong 1994), and some types of prey as far north as the Gulf of Alaska may have been depleted during the 1982–83 El Niño.

Some populations in the Far North may have actually benefited from warmer El Niño water (sea surface temperature rose up to 5°C in the Gulf of Alaska and 3°C in the southern Bering Sea). Juvenile northern fur seals from breeding populations in the Pribilof Islands had a 15% decrease in mortality during the El Niño, presumably from increased prey availability (York 1991). However, Gentry (1991) found no changes in mortality or behavior of the Pribilof populations of northern fur seals during the El Niño breeding season.

Effects of changes in prey abundance on cetaceans are poorly documented. Manzanilla (1989) found a hypocalcified dentine deposit in dusky dolphins from Peru, presumably laid down in conjunction with decreased prey abundance during the 1982–83 El Niño. Some species may follow their food supply to colder waters rather than suffer from an inadequate prey base. Cetaceans are very mobile and are not tied to specific land sites for breeding or hauling out.

The influence of the 1982–83 El Niño on marine mammal food supply varied spatially, in conjunction with observed effects on pinnipeds and cetaceans. Zooplankton stocks were reduced from Baja California (Mexico) to British Columbia, but may have increased in the Gulf of Alaska (Arntz et al. 1991). In the far North Pacific, prey species responded in different ways. Pacific mackerel (*Scomber japonicus*) and market squid (*Loligo opalescens*) migrated north and were found in greater numbers off of Oregon, Washington, and British Columbia. Hake (*Merlucius bilinearis*), pollock (*Pollachius chalcogrammus*), and Pacific herring (*Clupea harengus*) did not change significantly in distribution or abundance.

Pinniped and, to some extent, cetacean populations are strongly tied to their food supply, more so than to direct effects on breeding areas and thermoregulatory requirements. If prey changes in response to a long-term, uniform temperature increase, marine mammals in the far North Pacific may benefit, at least in the short term. Evidence based on short-term thermal phenomena suggests that more southerly prey species will move north into cooler waters and that more northerly species will be unaffected.

Polar bears appear to be adaptable in both temperature tolerance and diet, as evidenced by their ability to thrive in zoological displays in warm temperate climates and to switch from seal prey to feeding in garbage dumps in the wild (Stirling 1988). The biggest threat to polar bears in the wild is a decline in their pinniped prey (ringed and bearded seals), which might be affected by a decrease in the amount of sea ice available for breeding and hauling. Productivity in the polar seas is enhanced at the ice edge, so productivity in these areas could decline with less ice formation (Fittner 1993). The ringed seal is the second most abundant pinniped, with population estimates of six to seven million (Riedman 1990). A less productive Arctic would surely impact this and other marine mammals that rely on the high productivity of northern waters.

Summary

The combination of long-term warming, short-term fluctuations (e.g., those precipitated by El Niños), and habitat destruction (including overfishing) are more devastating than any one factor alone. The Steller sea lion provides a good example of this. Although population decline in the southern parts of its range may be partially due to a warmer climate, large breeding aggregations in the last 50 years, as far south as the Santa Barbara Channel Islands, suggest that sea lions are probably not heat-stressed in the northern parts of their range. Some breeding colonies in the north have undergone declines of approximately 80%, similar to those observed on Año Nuevo Island in the south (Loughlin et al. 1992). The population crash of sea lions in the northern portion of the range may have been caused by a decline in available prey due to competition with human fisheries for pollock, the Steller sea lion's principal prey (NMFS 1992). Further decreases in pollock biomass in conjunction with lower productivity due to long-term climatic warming would only make this competition more intense. A warmer climate will most likely impact pinniped and polar bear populations in the Arctic regions of the far North Pacific by decreasing the amount of available ice for breeding and hauling out. All marine mammal populations are likely to be negatively impacted by declines in available prey, except perhaps those residing in coastal zones.

What type of resource are marine mammals, and what effect would their decline or disappearance have on human populations in the North Pacific? If the term *resource* is defined as property or assets that can be utilized to obtain money (or food), then marine mammals have clearly been a highly exploited resource in the past. Of the 34 species listed in Table 7.1, only three species have not been exploited by commercial fisheries or used by Natives for subsistence. Although marine mammal commercial operations still exist in Russia and Japan, they no longer occur in the United States and Canada. From a purely economic standpoint, marine mammals are not presently a useful commercial resource, and many species may even compete with fisheries for prey. However, marine mammals continue to be important for subsistence hunters throughout Alaska and northern

Canada. In addition, the use of marine mammals as an economic resource for tourism has not been fully exploited in the far North Pacific, because many species are wary of humans and would be disturbed by groups of tourists, while others are widely dispersed and difficult (and expensive, in terms of travel) to encounter.

Species should be preserved because they are an integral part of ecosystems, not just because they are useful to humans. Marine mammals of the North Pacific, including those that migrate thousands of kilometers to feed in its productive waters, are an important biological resource. Along with the Steller sea lion, there have been population declines in harbor seals and northern fur seals due to unknown causes in the Gulf of Alaska (NMFS 1992). Climatic warming in conjunction with the overuse of fisheries by humans could result in the extinction of some marine mammal species, many of which have not recovered from past exploitation. The biological wealth represented by marine mammals in the North Pacific is a valuable resource for all humanity.

References

Arntz, W., W. G. Pearcy, and F. Trillmich. 1991. Biological consequences of the 1982–83 El Niño in the Eastern Pacific. In *Pinnipeds and El Niño: Responses to environmental stress*, edited by F. Trillmich and K. A. Ono, 22–42. Heidelberg: Springer-Verlag.

Costa, D. P., G. A. Antonelis, and R. L. DeLong. 1991. Effects of El Niño on the foraging energetics of the California sea lion. In *Pinnipeds and El Niño: Responses to environmental stress.*, edited by F. Trillmich and K. A. Ono, 156–165. Heidelberg: Springer-Verlag.

DeLong, R. L., and G. A. Antonelis. 1991. Impact of the 1982–1983 El Niño on the northern fur seal population at San Miguel Island, California. In *Pinnipeds and El Niño: Responses to environmental stress*, edited by F. Trillmich and K. A. Ono, 75–83. Heidelberg: Springer-Verlag.

DeLong, R. L., G. A. Antonelis, C. W. Oliver, B. S. Stewart, M. S. Lowry, and P. K. Yochem. 1991. Effects of the 1982–1983 El Niño on several population parameters and diet of California sea lions on the California Channel Islands. In *Pinnipeds and El Niño: Responses to environmental stress*, edited by F. Trillmich and K. A. Ono, 166–172. Heidelberg: Springer-Verlag.

Estes, J. A. 1991. Adaptations for aquatic living in carnivores. In *Carnivore behavior, ecology, and evolution*, edited by J. Gittleman, 242–282. Ithaca, NY: Cornell University Press.

Evans, P. G H. 1987. *The natural history of whales and dolphins*. New York: Facts on File.

Fittner, G. A. 1993. Impacts of climate change on living aquatic resources of the world. In *Impacts of climate change on resource management in the North*. Department of Geography Occasional Paper 16, edited by G. Wall, 99–125. Waterloo, Ontario: University of Waterloo.

Folland, C. K., T. Karl, and K. Y. A. Vinnikov. 1990. Observed climate variations and change. In *Climate change: The IPCC scientific assessment*, edited by J. T. Houghton, G. J. Jenkins, and J. J. Ephraums, 195–238. Cambridge: University of Cambridge.

Francis, J. M., and C. B. Heath. 1991. Population abundance, pup mortality, and copulation frequency in the California sea lion in relation to the 1983 El Niño on San Nicolas Island. In *Pinnipeds and El Niño: Responses to environmental stress*, edited by. F. Trillmich and K. A. Ono, 119–128. Heidelberg: Springer-Verlag.

Gentry, R. L. 1991. El Niño effects on adult northern fur seal at the Pribilof Islands. In *Pinnipeds and El Niño: Responses to environmental stress*, edited by F. Trillmich and K. A. Ono, 84–93. Heidelberg: Springer-Verlag.

Guerra, C. G., and G. Portflitt K. 1991. El Niño effects on pinnipeds in northern Chile. In *Pinnipeds and El Niño: Responses to environmental stress*, edited by F. Trillmich and K. A. Ono, 47–54. Heidelberg: Springer-Verlag.

Huber, H. R. 1991. Changes in the distribution of California sea lions north of the breeding rookeries during the 1982–83 El Niño. In *Pinnipeds and El Niño: Responses to environmental stress*, edited by F. Trillmich and K. A. Ono, 129–137. Heidelberg: Springer-Verlag.

Huber, H. R., C. Beckham, and J. Nisbet. 1991. Effects of the 1982–83 El Niño on northern elephant seals on the South Farallon Islands, California. In *Pinnipeds and El Niño: Responses to environmental stress*, edited by F. Trillmich and K. A. Ono, 219–233. Heidelberg: Springer-Verlag.

Leatherwood, S., and R. R. Reeves. 1978. Porpoises and dolphins. In *The marine mammals of eastern North Pacific and Arctic waters*, edited by D. Haley, 96–111. Seattle: Pacific Search Press.

Leatherwood, S., and W. A. Walker. 1979. The northern right whale dolphin, *Lissodelphis borealis* Peale, in the eastern North Pacific. In *Behavior of marine mammals*, vol. 3, *Natural history of cetaceans*, edited by H. E. Winn and B. L. Olla, 206–218. Heidelberg: Springer-Verlag.

Leatherwood, S., R. R. Reeves, and L. Foster. 1983. *The Sierra Club handbook of whales and dolphins*. San Francisco: Sierra Club Books.

Leatherwood, S., W. F. Perrin, V. L.'Kirby, C. L. Hubbs, and M. Dahlheim. 1980. Distribution and movements of Risso's Dolphin, *Grampus griseus*, in the Eastern North Pacific. *Fishery Bulletin* 77:951–963.

LeBoeuf, B. J., and J. Reiter. 1991. Biological effects associated with El Niño, Southern Oscillation 1982–83, on Northern elephant seals breeding at Año Nuevo, California. In *Pinnipeds and El Niño: Responses to environmental stress*, edited by F. Trillmich and K. A. Ono, 206–218. Heidelberg: Springer-Verlag.

LeBoeuf, B. J., K. A. Ono, and J. Reiter. 1991. History of the Steller sea lion population at Año Nuevo Island, 1961–1991. *NMFS Southwest Fisheries Science Center Administrative Report* LJ–91–45C.

Loughlin, T., A. S. Perlov, and V. A. Vladimirov. 1992. Range-wide survey and estimation of total number of Steller sea lions in 1989. *Marine Mammal Science* 8:220–239.

Manzanilla, S. R. 1989. The 1982–1983 El Niño event recorded in dentinal growth layers in teeth of Peruvian dusky dolphins (*Lagenorhynchus obscurus*). *Canadian Journal of Zoology* 67:2120–2125.

National Marine Fisheries Service (NMFS). 1992. *Recovery plan for the Steller sea lion (*Eumetopias jubatus*).* Report prepared by the Steller Sea Lion Recovery Team. Silver Spring, MD: National Marine Fisheries Service.

Ono, K. A., D. J. Boness, and O. T. Oftedal. 1987. The effect of a natural environmental disturbance on maternal investment and pup behavior in the California sea lion. *Behavioral and Ecological Sociobiology* 21:109–118.

Ono, K. A., W. Hood, and K. Anderson. California sea lion pups on Año Nuevo Island: incipient colony or environmentally induced accidents? *Marine Mammal Science.* In review.

Quinn, W. H., V. T. Neal, and S. E. Antuñez de Mayolo. 1987. El Niño occurrences over the past four and a half centuries. *Journal of Geophysical Research* 92C:14449–14461.

Riedman, M. 1990. *The pinnipeds: Seals, sea lions and walruses.* Berkeley: University of California Press.

Royer, T. C. 1985. Coastal temperature and salinity anomalies in the northern Gulf of Alaska, 1970–84. In *El Niño north,* edited by W. S. Wooster and D. L. Fluharty, 107–115. Seattle: Washington Sea Grant Program, University of Washington.

Stewart, B. S., and P. K. Yochem. 1991. Northern elephant seals on the southern California Channel Islands and El Niño. In *Pinnipeds and El Niño: Responses to environmental stress,* edited by F. Trillmich and K. A. Ono, 234–243. Heidelberg: Springer-Verlag.

Stewart, B. S., and R. L. DeLong. 1994. Postbreeding foraging migrations of northern elephant seals. In *Elephant seals: Population, ecology, behavior, and physiology,* edited by B. J. LeBoeuf and R. M. Laws, 290–309. Berkeley: University of California Press.

Stirling, I. 1988. *Polar bears.* Ann Arbor: University of Michigan Press.

Trillmich, F., K. A. Ono, D. P. Costa, R. L. DeLong, S. D. Feldkamp, J. M. Francis, R. L. Gentry, C. B. Heath, B. J. LeBoeuf, P. Majluf, and A. E. York. 1991. In *Pinnipeds and El Niño: Responses to environmental stress,* edited by F. Trillmich and K. A. Ono, 247–270. Heidelberg: Springer-Verlag.

VanBlaricom, G. R., and J. A. Estes. 1988. *The community ecology of sea otters.* Heidelberg: Springer-Verlag.

Wells, R. S., L. J. Hansen, A. Baldridge, T. P. Dohl, D. L. Kelly, and R. H. Defran. 1990. Northward extension of the range of bottlenose dolphins along the California coast. In *The bottlenose dolphin,* edited by S. Leatherwood and R. R. Reeves, 421–431. New York: Academic Press.

Wynne, K., and P. Folkens. 1993. *Guide to marine mammals of Alaska.* Fairbanks: Alaska Sea Grant College Program, University of Alaska.

York, A. E. 1991. Sea surface temperatures and their relationship to the survival of juvenile male northern fur seals from the Pribilof Islands. In *Pinnipeds and El Niño: Responses to environmental stress,* edited by F. Trillmich and K. A. Ono, 94–105. Heidelberg: Springer-Verlag.

8

Response of anadromous fish to climate change in the North Pacific

Richard J. Beamish

Natural selection occurs when there are extreme fluctuations in the environment, indicating that variability and change are common to all living things. However, in fisheries science and fisheries management, we find ourselves generally unprepared for changes in the environment. As world fisheries developed from the 1950s to the 1970s, there was an urgent need to determine how much fish could be harvested. High priority was given to research that would lead to the development of methods to identify levels of catch that were sustainable and hopefully stabilize the developing industries and economies of maritime communities. Research on the relationships between marine environments and fish populations received less attention. Consequently, priorities for the development of techniques to assess the impacts of fishing limited our understanding of the effects of the environment, which tended to be modeled as random events.

Several recent articles have used the results of general circulation models to forecast the impacts of a changing climate on freshwater and marine fishes. I will not attempt to review all of these studies or to add my speculations to this extensive literature. Rather, I will emphasize the importance of understanding the impact of past climatic events on populations of several important fish species of the North Pacific. This will allow us to anticipate the potential consequences of our unprecedented alteration of the atmosphere.

If we are to draw inferences from past events to interpret the future, the continuing debate concerning the relative effects of climate/environment versus fishing effects on the dynamics of fish populations must be resolved. Resolving this issue will not decide who is right and who is wrong, and it should not reflect criticism of fisheries management. Rather, it will increase our understanding of the parameters that regulate the abundance of fishes.

The relative importance of certain environmental factors and fishing effects in the dynamics of fish populations is widely debated in scientific circles as well as in the media. It is worth reviewing a few examples of this controversy because we cannot assess the impacts of climate change unless the impact of climate on fish populations can be documented.

Controversies over climatic impacts on fish population dynamics

> I think the tension in Newfoundland caused by the crisis in the northern cod stock might be relieved a little if it were publicly realized that part of the trouble may be due to Nature and not to Man. Those concerned with the fishery, scientists as well as politicians and others, seem to have forgotten the story of the West Greenland cod fishery. (M. J. Dunbar, Toronto *Globe and Mail;* July 4, 1993, p. A17)

A classic story in fisheries management is that of the West Greenland cod (*Gadus callarias*) (Cushing 1982). The West Greenland cod fishery increased suddenly in the 1930s. Another dramatic increase occurred from the late 1950s to the late 1960s with catches ranging from 300,000 to 450,000 metric tons. The stocks collapsed beginning in the late 1960s. The fluctuations in abundance were not caused by fishing practices, but by changes in ocean temperature associated with wind patterns.

In one of the first environment versus fishing controversies, Burkenroad (1948) reviewed the factors affecting the Pacific halibut (*Hippoglossus stenolepis*) population and concluded that a decrease in abundance before 1930 was not the exclusive result of overfishing and that increases in abundance after 1930 were related to improved ocean survival. Burkenroad stated that fish populations appear to fluctuate naturally: the population declines to its lowest level and then begins to increase in abundance thereafter. When a scarcity occurs, it is a stimulus to study the fishery and regulate it, so regulation generally coincides with low points in natural periodicities. Therefore, because one would expect to find increases in abundance following regulation, even if fishing had nothing to do with the scarcity, an increase in abundance is not itself sufficient evidence for a causal connection between regulation and rising fish populations.

The International Pacific Halibut Commission (IPHC) was established in 1923 to manage the declining halibut population fished by Canada and the United States. Burkenroad (1948) said that governments are best at taking action when there is a crisis, and in fisheries a crisis occurs when stocks are at a natural low point. Action taken when abundance is low may be successful because the stocks are about to increase anyway. One can imagine how the IPHC viewed the conclusion that its efforts were not the principal reasons for the stock increases! IPHC Director W. F. Thompson rejected Burkenroad's arguments by stating that the fishery was responsible for the condition of the stock. Other scientists joined the debate, although supporters of "fishery impacts" claimed victory.

One lesson from the Thompson-Burkenroad debate is that managing fisheries in a changing climate requires a process of decision making that is not prompted by crisis. Another is that in the early 1980s, the IPHC adopted the position that the abundance of halibut fluctuates naturally (Figure 8.1), and that the fishery should take a constant percentage of the exploitable biomass, ensuring that reproduction is sufficient to replace the present population. The IPHC now recognizes that the effects of both fishing and the ocean environment must be considered when setting quota allocations.

The 1976–77 event in the mid-Pacific

Variability in climate is common on an interannual and interdecadal scale, and there have been gradual and abrupt shifts to the decadal scale during this century (Chapter 2). A striking climatic shift occurred in the North Pacific Ocean in the winter of 1976–77 (Beamish 1993; Polovina et al. 1994), the result of atmospheric anomalies recorded several months earlier (Miller et al. 1994). The effect on the ocean was basinwide, resulting in altered patterns of heat flux and ocean current advection. Intensification of the Aleutian Low Pressure System (a region of low pressure over the Gulf of Alaska) resulted in the lowest pressure measured since 1940–41, a drop of 2 millibars (mb) over the entire North Pacific. The cen-

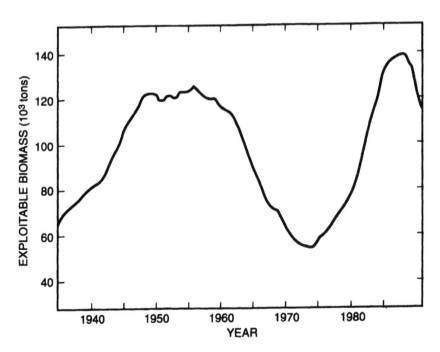

Figure 8.1. The exploitable biomass of Pacific halibut from 1935–1991. Exploitable biomass is an estimate of the total weight of the halibut population that is available to the commercial fishery. In recent years the exploitable biomass is declining for natural reasons. Data courtesy of the International Pacific Halibut Commission.

ter of the Aleutian Low moved eastward and intensified by an average of 4.3 mb
for November through March (7–9 mb in January alone).

Ebbesmeyer et al. (1991) collected data for the period 1968–84 on 40 atmospheric, ocean, and biological variables; most data were from the North Pacific
and adjacent coastal areas, but some were from the South Pacific. The time series
of data were normalized by subtracting the mean of the time series and dividing
by the standard deviations, then averaging each dimensionless variable for each
year. After 1976–77, 22 variables increased, and 18 decreased. When the
researchers reversed the signs of the variables that decreased, the resulting single
time series had a steplike shift in the 1976–77 winter, which accounted for 89% of
the variance in the component time series. The study clearly showed that there
was a synchronous response in a diverse set of biological and physical time series
over a vast area.

The abrupt climate change in 1976–77 was associated with a reversal of a
declining trend in Pacific salmon (Beamish and Bouillon 1993; Hare and Francis
1995) and with strong production of non-salmon species (Hollowed and Wooster
1992; McFarlane and Beamish 1992; Beamish 1993; Beamish and Bouillon
1995). There are five common species of Pacific salmon in the commercial fishery. Approximately 89% of the commercial catch consists of pink (*Oncorhynchus
gorbuscha*), chum (*O. keta*), and sockeye (*O. nerka*) salmon; the other two
species, coho (*O. kisutch*) and chinook (*O. tshawytscha*), represent 11% of the
total catch. The all-nation catches of pink, chum, and sockeye have fluctuated
between 275,600 and 856,500 metric tons between 1925 (when reliable data
became available) and 1993, with the lowest catch in 1974. By adding the catches
together for each species, excluding coho and chinook, patterns of catches were
remarkably similar (Beamish and Bouillon 1993). The synchronous pattern of
catch trends indicates that some large-scale event in the marine environment
influenced salmon production.

The Aleutian Low Pressure System was the mechanism most likely associated with increased ocean productivity (Figure 8.2). During periods of increasing
pressure, strong mid-ocean upwelling is believed to increase productivity, and the
associated horizontal divergence would transport nutrients and plankton into
coastal areas. Increased productivity resulting from increased surface mixing of
deep ocean water may have been responsible for the improved overall production
in the North Pacific in the 1980s (Miller et al. 1994; Polovina et al. 1994). Plankton production is positively correlated with fish production (McFarlane and
Beamish 1992; Beamish and Bouillon 1993), and there was a general increase in
plankton during the influx through Aleutian Lows in the 1980s (Brodeur and
Ware 1992). Therefore, the increases in plankton resulting from intensification of
the Aleutian Low were associated with improved fish production in the North
Pacific.

Historical shifts in Alaskan salmon production similar to that observed in
1976–77 may also have been driven by climate. For example, Hare and Francis

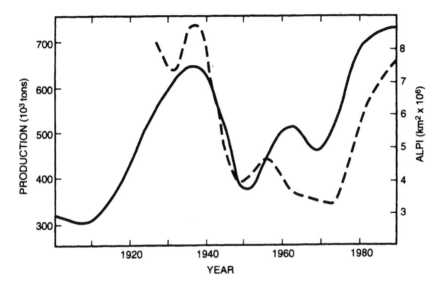

Figure 8.2. Comparison of the combined all-nation catch of pink, chum, and sockeye salmon (bro-ken line) and the smoothed Aleutian Low Pressure Index (solid line). Adapted from Beamish and Bouillon (1993), figure 7. Reprinted with permission of the National Research Council of Canada.

(1995) identified a less dramatic and negative change for pink and sockeye stocks in the mid-1950s. Alaska salmon populations appear to alternate between high and low production with smaller scale variability within these regimes. The shift from one regime to another occurs quickly, emphasizing the need to understand the causes of changes in catch trends. If the decreases in catch result from over-fishing, proper application of fishing theory will enable the stocks to rebuild. If the declines are associated with climate or other environmental factors, reduc-tions in catch will not necessarily result in increased fish abundance until the nat-ural trend is reversed.

The climatic event of 1976–77 also affected non-salmon stocks (Hollowed and Wooster 1992; Beamish 1993). Hollowed and Wooster (1992) identified two environmental states in the northeastern Pacific, one associated with weak circu-lation and the other with strong circulation. As the states changed, there was an alternating pattern in coastal sea surface temperatures. Sea surface temperatures were cool from 1965 to 1976 (except for 1970), and warm from 1977 to 1987. Synchronous strong year classes occurred during the warm periods and during periods of strong circulation. The change from cool to warm coastal sea surface temperatures in 1976–77 (Hollowed and Wooster 1992) was associated with a general increase in the number of strong year classes, that is, a year of exceptional survival of larval fish.

During 1977 and 1978, there was also exceptionally strong year class pro-duction of 12 non-salmon species, with 1977 being the strongest since the early

1950s (Beamish 1993; Beamish and Bouillon 1993). In 1977, 10 of the 12 species examined had positive anomalies, the largest number in the 27–year time series from 1960 to 1986.

Other areas and species in the mid-Pacific were also affected by climatic changes in the 1970s and 1980s. For example, there was a strong linkage among the atmosphere, ocean mixing, and production of a wide variety of marine taxa in the Hawaiian Archipelago (Polovina et al. 1994). More frequent deep-mixing events during this period resulted in increased populations of reef fishes, flying fishes (e.g., *Cypselurus californicus*), squid (*Loligo opalescens*) and spiny lobster (*Panulirus interruptus*), as well as predatory sea birds and monk seals (*Monachus tropicalis*). The regime shifted from the mid- to late 1980s, with fewer deep-mixing events and synchronous declines in the survival of the species just mentioned.

At the time of the 1976–77 event, some coastal states extended their exclusive coastal fishing zones to 200 nautical miles. It was presumed that this change would allow better management and optimization of yields—in other words, more fish. It is not surprising that the major increase in ocean productivity in the North Pacific at this time would be linked to the new management regimes and not to climate.

The 1976–77 event in coastal areas

The global scale climatic events that improved mid-ocean productivity were associated with declines in the survival of some Pacific salmon species in coastal areas. There was a dramatic change in coho production off the coasts of California, Oregon, and southern Washington, the so-called Oregon Production Index (OPI) area. Since the turn of the century, coho have been raised in hatcheries in increasing numbers to maintain abundance as wild coho stocks declined (Pearcy 1992). By the early 1970s, hatcheries were producing 30–35 million smolts (young salmon stage capable of migrating from fresh water to the ocean). Adult returns fluctuated, but were considered to be high. In 1977, adult returns suddenly crashed and remained generally low into the 1980s, even though an increasing number of smolts were released. Recent adult returns have been so low that severe restrictions have been placed on recreational and commercial fisheries.

Explanations for these declines range from poor hatchery practices to changes in the ocean environment (Pearcy 1992). An association with the ocean environment is, in fact, apparent. Survival of hatchery-reared smolts in strong upwelling years was almost twice the survival in weak years (Nickelson 1986). Warmer temperatures were correlated with lower survival in upwelling years, but there was no apparent relationship in weak upwelling years. In 1976, coastal sea surface temperatures changed from cool to a period of warm years (1976–87), in conjunction with changes in sea surface temperatures and intensification of the

Aleutian Low. Sea surface temperatures off the coasts of Washington and Oregon increased as cyclonic circulation pushed more warm water south and along the coasts. Although the exact nature of the interactions is unclear, it is believed that changes in the ocean environment affected predator-prey relationships and consequently affected the marine survival of coho.

Changes in the ocean that occurred off the Oregon coast occurred at the same time as changes near inshore areas such as Puget Sound (Ebbesmeyer et al. 1989). As the center of the Aleutian Low moved away from the west coast of North America, storms in the Gulf of Alaska became less frequent, and snow deposits in Puget Sound mountain watersheds decreased. Fresh water discharge into Puget Sound decreased, changing the pattern of bottom water flow into the sound. The pattern was reversed when the Aleutian Low moved toward North America.

In the Strait of Georgia, between Vancouver Island and mainland British Columbia, bottom water temperature increased approximately 0.5°C beginning in 1977 (Beamish 1993). The change in bottom water temperature was synchronous with a declining trend in annual Fraser River discharge. Total catches of chinook salmon in Strait of Georgia fisheries averaged 284,000 from 1952 to 1969 (Figure 8.3). Beginning in 1970, catches increased to a maximum of 775,000 in 1976, averaging 755,000 from 1976 to 1978. Beginning in 1979, catches declined steadily to 175,000 in 1987, when catch levels were regulated. Catches averaged 167,000 between 1987 and 1991. A decline in chinook salmon in the Strait of

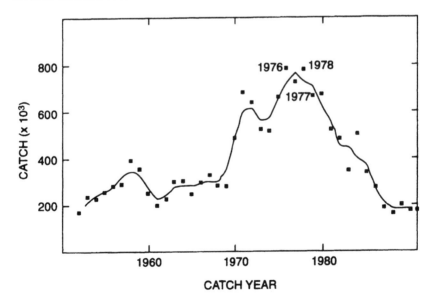

Figure 8.3. Total catch of chinook salmon in the Strait of Georgia sport and commercial fisheries from 1955–1991 (solid squares). A 3–year running average of the total catch shows a decline in catch beginning in 1979.

Georgia was synchronous with the decline of coho in coastal Oregon and Washington. The decline in survival of hatchery-reared coho and chinook salmon (Beamish et al. 1995) was synchronous with the Strait of Georgia decline in chinook salmon.

Chinook and coho survival from smolt to adult for all hatcheries was high for brood years from 1973 to 1976 and for fish going to sea from 1974 to 1977 (Figure 8.4; Beamish et al. 1995). Beginning with the fish that went to sea in the late

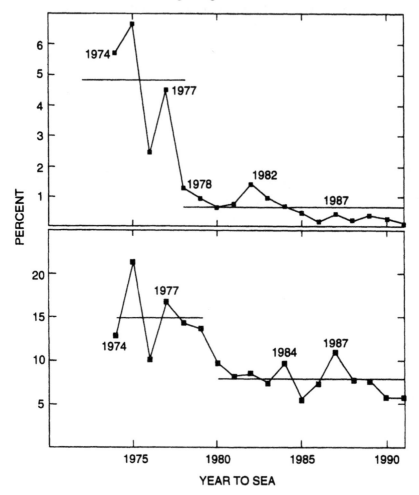

Figure 8.4. Marine survival percentage of chinook (upper) and coho (lower) salmon in the Strait of Georgia that were released from hatcheries near the Fraser River and along the Strait of Georgia. The years indicate when the young salmon entered salt water. The average survival for chinook salmon (upper, solid trend lines) was 4.8% from 1974 to 1977 and 0.7% from 1978 to 1991. Average survival for coho salmon (lower, solid trend lines) was 14.9% from 1974 to 1979 and 8.0% from 1980 to 1991.

1970s, there was a dramatic and abrupt decline in survival of hatchery-reared chinook and coho; survival has remained low since then.

There is no direct proof that the abrupt changes in the marine environment of the Strait of Georgia and Puget Sound are linked to the changes in chinook catch and chinook and coho hatchery survivals. However, the synchrony of the events, including declines in coho catches off the coast of Oregon, provides strong circumstantial evidence that these changes were related to the climatic events. It is impressive to see how even short-term climatic phenomena can have dramatic and opposing impacts on salmon stocks.

The impact of climate change on salmon in fresh water

Changes that could occur in fresh water as a consequence of climate change would also have a major impact on Pacific salmon: migrations, spawning, hatching, and early rearing phases of the life history of salmon would be affected. Physical changes in temperature, precipitation, groundwater discharge, and increased ice-free periods for lakes could affect community structure and the survival, growth, and distribution of salmon species (Meisner et al. 1987; Glantz 1990; Magnuson et al. 1990; Chatters et al. 1991; Neitzel et al. 1991; Northcote 1992).

One approach to stewardship of fisheries and aquatic resources in times of climate change is to identify expected changes for species and to then propose managerial responses. This approach would focus attention on the issue, identify areas requiring more information, and commit agencies and institutions to a dialogue with all stakeholders, including subsistence fishermen.

The Fraser River drainage in British Columbia is a major producer of Pacific salmon, accounting for 30 to 40% of all Pacific salmon produced in Canada. Because numerous stocks of the five species of salmon are at or near the southern limit of their range, the early impacts of climate change should be detectable in these stocks. Levy (1992) investigates these potential impacts by formulating eight hypotheses (Table 8.1) and considering the possible responses of salmon to climate change.

Hypothesis 1 (Table 8.1) is highly probable because there is a direct relationship between increased river temperatures and prespawning mortalities for all salmon. The early migrations would be particularly vulnerable. While hypothesis 2 is possible, the impacts of climatic warming on winter water temperatures are uncertain. Winter water temperatures relate to groundwater base flows, lake water runoff, precipitation levels, and perhaps changes in snowmelt patterns. All these factors can affect water temperature over spawning grounds. Hypothesis 3 is highly probable within the coastal portion of the Fraser River watershed, where the greatest increases in winter stream discharge are expected. Hypothesis 4 is particularly relevant for coho and chinook salmon, which remain in fresh water

Table 8.1. Hypotheses proposed by Levy (1992) as possible consequences of climate change on Pacific salmon stocks in the Fraser River.

Hypothesis 1: A warmer climate will increase water temperatures and decrease flows during spawning migrations, increasing prespawning mortality and reducing egg deposition.

Hypothesis 2: A warmer climate will increase water temperatures during egg incubation stages, causing premature fry emergence and increased fry-to-smolt mortality.

Hypothesis 3: A warmer climate will increase the severity and frequency of winter floods, thereby reducing egg-to-fry survival rates.

Hypothesis 4: A warmer climate will increase stream and river water temperatures away from the thermal optima for salmon, creating suboptimal salmon habitat conditions in the Fraser River.

Hypothesis 5: A warmer climate will alter the timing and volume of stream flow discharges, reducing the capability of streams to produce juvenile salmon.

Hypothesis 6: A warmer climate will cause altered limnological conditions in lakes in the Fraser River drainage, thereby reducing their suitability as nursery habitats for juvenile sockeye salmon.

Hypothesis 7: A warmer climate will increase water temperatures, resulting in altered aquatic community structure, adversely affecting salmon populations in the Fraser River.

Hypothesis 8: A warmer climate will shift the timing of spring freshet, effectively increasing the mortality rate of salmon smolts as they migrate down the river.

during their first year after hatching. Warm summer water temperatures may be too high for optimal growth and may force the young salmon into suboptimal habitats. The impacts of poor growth on the remaining life history stages are unknown, but are probably unfavorable. Coho salmon are the most vulnerable to warming, because their embryonic development is sensitive to warm winter temperatures (Beacham and Murray 1990). Warm water during development could reduce size at emergence and affect subsequent survival.

Hypothesis 5 points out another consideration with respect to habitat for juvenile salmon. Stream flow discharge patterns have a high degree of variation in any case, and changes in the variation of timing and volume could create unsuitable flow conditions on a more frequent basis. Hypothesis 6 identifies potentially serious problems associated with the early life history of sockeye salmon. Virtually all sockeye stocks spawn in rivers that flow into or out of lakes. After hatching, fry move into the lake, where they must remain for a minimum of one year. Although there is uncertainty about how lakes will respond to a warmer climate, we do know that sockeye salmon in the Fraser River are at the extreme southern limit of their distributions and will be sensitive to changes in limnological conditions.

Hypothesis 7 indicates that increased water temperature would change aquatic community structure into a warmer water community with different functions and processes. The specific factors that regulate salmon abundance are not clearly identified, making it difficult to predict the impacts of altered ecosystems.

Hypothesis 8 suggests that earlier migrations down and out of the river are possible, but there is no evidence that there would be increased mortality during this earlier migration (Levy 1992). An increase in early marine mortality is also a possibility. The timing of spring floods is a key parameter in modeling the impacts of climatic warming on chinook salmon survival in the Columbia River (Chatters et al. 1991).

Salmon at their northern limits

Salmon are particularly susceptible to temperature fluctuations because they have adapted to thermal regimes in both fresh and salt water. At the northern limits of salmon distribution, projected climatic changes would warm both marine and freshwater habitats, especially in the winter. High temperature has a profound effect on fishes because they cannot regulate their body temperature. Extreme temperatures may kill eggs, juveniles, or adult fish; less extreme temperatures can affect growth, reproduction, and movement. Recommended temperatures for most Pacific salmon in fresh water range from about 7° to 16°C (Reiser and Bjornn 1979), with extremes from 3° to 20°C. Upper lethal temperatures are 25° to 26°C. Southern rivers could approach these higher limits under projected climatic scenarios. In the marine phase of the life history of salmon, there are critical temperatures of 9°–10°C that restrict feeding areas for salmon to areas cooler than these temperatures (Welch et al. 1995). Ocean warming, particularly in the winter, could favor northern areas for rearing of salmon.

Warming of fresh water in the North may also improve production. Much of the increases in Pacific salmon abundance in the 1980s occurred in Alaska stocks, possibly indicating that warming in fresh water and coastal areas at this time was beneficial for salmon production. However, the function of northern aquatic systems has not been well documented, and large temperature increases could have unforeseen effects on Pacific salmon survival.

Levy's (1992) hypotheses (Table 8.1) and evaluation of linkages between climate and biological systems could be used to plan for managing fragile northern freshwater and marine ecosystems in a warmer climate. We have community plans for other natural (e.g., earthquakes) and human-caused (e.g., oil spills) phenomena. We should also develop plans for the impacts of climate change on salmon fisheries throughout the range of Pacific salmon.

Fisheries management and research in a changing climate

Fisheries management attempts to control the abundance of species that humans want to exploit. Should fisheries management be a stewardship of a resource as it responds to natural events? If so, management can be defined as a process of removing yield, while ensuring that we do not inhibit a species from sustaining itself. Past experience indicates that we may need to become more sensitive to the possibility that we cannot control the abundance of fish for extended periods.

Will we respond to environmental changes to maintain fisheries and species compositions, or will we adapt to the changes by anticipating some end point to the change? Changes in distributions, including extinctions and natural introductions from invasions, would be important consequences (Tonn 1990). However, by the time changes become obvious, priorities for water use may not include the sustainability of fish populations (Neitzel et al. 1991) .

Summary

Climate change will affect water temperature, sea level, wetlands, estuaries, river discharge, ocean currents, and plankton. Physical changes may be more serious than those that occurred in 1976–77. Thus, we need to identify the signals of change and understand their causes faster than we did for the 1976–77 event. We may not be able to wait until science provides management with risk-free advice: Scientists and managers may have to make choices that are not easily evaluated in the short term rather than wait for the inevitable crisis.

Long-term fisheries research may be more difficult in the future because the instability of ecosystems will make it difficult to establish controls. If adaptive studies can become more opportunistic, then assessment of natural changes may yield valuable information. The kinds of information needed will be tempered not only by a changing climate, but by an expanding and changing human population. New issues may also arise. Will the role of salmon enhancement by hatcheries and other means change during periods of natural declines in salmon survival? What will happen to subsistence fisheries in the North and in the South? Past experience indicates that stewardship of our salmon resources through a change to a warmer climate will be both complex and challenging.

References

Beacham, T. D., and C. B. Murray. 1990. Temperature, egg size, and development of embryos and alevins of five species of Pacific salmon: A comparative analysis. *Transactions of the American Fisheries Society* 119:12–20.

Beamish, R. J. 1993. Climate and exceptional fish production off the west coast of North America. *Canadian Journal of Fisheries and Aquatic Science* 50:2270–2291.

Beamish, R. J., and D. R. Bouillon. 1993. Pacific salmon production trends in relation to climate. *Canadian Journal of Fisheries and Aquatic Science* 50:1002–1016.

Beamish, R. J., and D. R. Bouillon. 1995. Marine fish production trends off the Pacific coast of Canada and the United States. *Canadian Special Publication of Fisheries and Aquatic Science* 121:585–591.

Beamish, R. J., B. E. Riddell, C. M. Neville, and Z. Zhang. 1995. Evidence of a relationship between changes in chinook salmon catches in the Strait of Georgia and shifts in marine production. *Fisheries Oceanography.* In press.

Brodeur, R. D., and D. M. Ware. 1992. Long-term variability in zooplankton biomass in the subarctic Pacific Ocean. *Fisheries and Oceanography* 1:32–38.

Burkenroad, M. 1948. Fluctuations in the abundance of Pacific Halibut. *Bulletin of the Binghamton Oceanographic College* 11:81–129.

Chatters, J. C., D. A. Neitzel, M. J. Scott, and S. A. Shankle. 1991. Potential impacts of global climate change on Pacific Northwest spring chinook salmon (*Oncorhynchus tshawytscha*): An exploratory case study. *Northwest Environmental Journal* 7:71–92.

Cushing, D. H. 1982. *Climate and fisheries.* New York: Academic Press.

Dunbar, M. J. 1993. Why have the cod gone? *Toronto Globe and Mail,* Aug. 17, 1993, p. A17.

Ebbesmeyer, C. C., C. A. Coomes, G. A. Cannon, and D. E. Bretschneider. 1989. Linkage of ocean fjord dynamics at decadal period. *Geophysical Monographs* 55:399–417.

Ebbesmeyer, C. C., D. R. Cayan, D. R. McLain, F. H. Nichols, D. H. Peterson, and K. T. Redmond. 1991. 1976 step in the Pacific climate: Forty environmental changes between 1968–1975 and 1977–1984. In *Proceedings of the seventh annual Pacific climate workshop,* edited by J. L. Betancourt and V. L. Tharp, 129–141. Interagency Ecological Studies Program Technical Report 26. Sacramento: California Department of Water Resources.

Glantz, M. H. 1990. Does history have a future? Forecasting climate change effects on fisheries by analogy. *Fisheries* 15:39–44.

Hare, S. R. and R. C. Francis. 1995. Climate change and salmon production in the northwest Pacific Ocean. *Canadian Special Publication of Fisheries and Aquatic Science* 121:357–372.

Hollowed, A. B., and W. S. Wooster. 1992. Variability of winter ocean conditions and strong year classes of Northeast Pacific groundfish. *ICES Marine Science Symposium* 195:433–444.

Levy, D. A. 1992. Potential impacts of global warming on salmon production in the Fraser River watershed. *Canadian Technical Report of Fisheries and Aquatic Science* 1889:96 p.

Magnuson, J. J., J. D. Meisner, and D. K. Hill. 1990. Potential changes in the thermal habitat of Great Lakes fish after global warming. *Transactions of the American Fisheries Society* 119:254–264.

McFarlane, G., and R. J. Beamish. 1992. Climate influence linking copepod production with strong year-classes in sablefish (*Anoplopoma fimbria*). *Canadian Journal of Fisheries and Aquatic Science* 49:743–753.

Meisner, J. D., J. L. Goodier, H. A. Regier, B. J. Shuter, and W. J. Christie. 1987. An assessment of the effects of climate warming on Great Lakes Basin fishes. *Journal of Great Lakes Research* 13:340–352.

Miller, A. J., D. R. Cayan, T. P. Barnett, N. E. Graham, and J. M. Oberhuber. 1994. The 1976–77 climate shift of the Pacific Ocean. *Oceanography* 7:21–26.

Neitzel, D. A., M. J. Scott, S. A. Shankle, and J. C. Chatters. 1991. The effect of climate change on stream environments: The salmonid resource of the Columbia River Basin. *Northwest Environmental Journal* 7: 271–293.

Nickelson, T. E. 1986. Influences of upwelling, ocean temperature, and smolt abundance on marine survival of coho salmon (*Oncorhynchus kisutch*) in the Oregon Production Area. *Canadian Journal of Fisheries and Aquatic Science* 43:527–535.

Northcote, T. G. 1992. Prediction and assessment of potential effects of global environmental change on freshwater sportfish habitat in British Columbia. *Geojournal* 28:39–49.

Pearcy, W. G. 1992. *Ocean ecology of North Pacific salmonids*. Seattle: University of Washington Press.

Polovina, J. J., G. T. Mitchum, N. E. Graham, M. P. Craig, E. E. Demartini, and E. N. Flint. 1994. Physical and biological consequences of a climate event in the central North Pacific. *Fisheries Oceanography* 3:15–21.

Reiser, D. W., and T. C. Bjornn. 1979. Influence of forest and rangeland management on anadromous fish habitat in the western United States and Canada. 1. Habitat requirements of anadromous salmonids. *USDA Forest Service General Technical Report PNW–96*. Portland, OR: Pacific Northwest Research Station.

Tonn, W. M. 1990. Climate change and fish communities: A conceptual approach. *Transactions of the American Fisheries Society* 119:337–352.

Welch, D. W., A. I. Chigiensky, and Y. Ishida. 1995. Upper thermal limits on the oceanic distribution of Pacific Salmon (*Oncorhynchus* spp.) in the spring. *Canadian Journal of Fisheries and Aquatic Science* 52.

PART III

Human Populations and Natural Resources: Historical and Contemporary Perspectives

The previous sections have illustrated the uncertainties associated with identifying the potential impacts of rapid climate change on ecosystems and individual species. It is even more problematic to make inferences about the effects of changes in natural systems on components of social systems.

Our first reaction may be to retreat from this quandary, to withdraw to intellectually safe ground until certainty replaces conjecture. However, at risk are not only the unique natural systems of the Far North, but also the subsistence-based cultures of contemporary Native hunters and trappers and people of non-Native descent who live in rural northern communities. To the people of the rural Far North, these cultures may be as important as life itself. The existence of these cultures has great symbolic relevance for modern society as well.

Our examination of northern social systems begins with an overview of the history of human occupation of the Far North and the dynamic interaction between natural systems and indigenous human systems. Specific social and cultural elements at risk in northern communities are determined by examining the dependence of indigenous and other rural people on natural resources. We learn that substantial numbers of contemporary Northerners depend on locally harvested wild foods. But the land is more than a source of food. Hunting is not equivalent to visiting a grocery store in the city. The harvest and distribution of wild food permeates northern Native culture and often defines personal identity and self-esteem.

An issue is raised here that is increasingly relevant in debates concerning natural resource management policy worldwide (e.g., old-growth forest harvest, wolf kills, ivory sales, whale harvest, grazing on public lands). "Local ownership" of natural resources is increasingly being denied or challenged, especially (but not exclusively) when consumptive uses occur on public lands. Management claims are being made in the name of all citizens and perhaps even in the name of the global community. To what extent will external publics (whoever they are) gain control of the management of locally relevant natural resources in the Far

North? How will this external control affect protected areas if both natural and social systems are under stress?

The following discussions of the dynamic interaction between natural and social systems emphasizes that the future evolution of northern rural cultures cannot be defined by idealized views of the past or present. Rather, we need to examine what is socially at risk in rural areas if biophysical systems of the Far North are subjected to rapid change. Then we can attempt to conceptualize the potential impact of altered natural systems on rural communities. While human cultures are by definition characterized by persistence and continuity, northern social systems of the late twenty-first century may be very different than in the past or how they exist today.

9

Increments, ranges, and thresholds: Human population responses to climate change in northern Alaska

Steve J. Langdon

Climatic factors such as floods, storms, and extreme cold greatly affect human capacities to harvest and process animals used for food. Animals taken for food (e.g., salmon [*Oncorhynchus* spp.], whales [Cetacea], walrus [*Odobenus rosmarus*], caribou [*Rangifer tarandus*]) are generally available during relatively short periods in summer and early autumn; this is especially true for terrestrial wildlife (Chapter 6) and anadromous fish (Chapter 8) populations. Huge variations in abundance are possible, including near total collapse. Much of the year-to-year variation in wildlife populations can be attributed to variation in climatic factors.

If climate change occurs at large time and geographic scales in the future, we can expect even greater disruptions in the distribution and abundance of wildlife. This will clearly have an impact on wildlife resources for economic and cultural sustenance. Mainland Alaska north of the Yukon River, including St. Lawrence and Little Diomede islands, contains considerable variation in climate, biota, and human cultures. This chapter explores the relationship of human populations to climatic and environmental conditions in this region.

Indigenous groups: Traditional resource use and settlement

Indigenous groups

Three Native groups defined by their language—Inupiat, Yup'ik, and Athabascan—have occupied this area since the nineteenth century (Figure 9.1). The Inupiat and Yup'ik are often referred to as Eskimo—their languages belong to the same linguistic family (Eskaleutian); populations speaking these languages appear to be closely related genetically (Szamarthy 1984). Athabascan groups

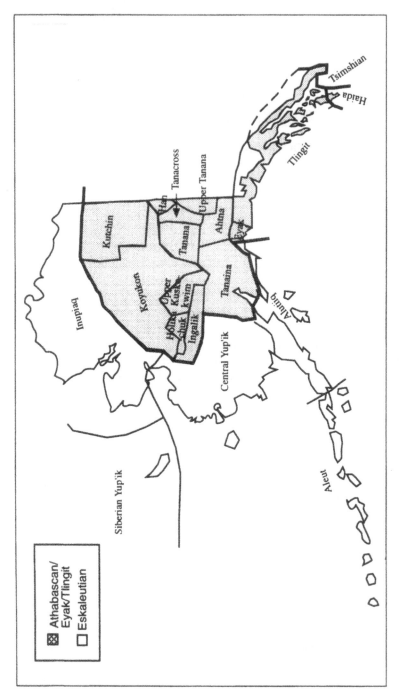

Figure 9.1. *Major linguistic groups of indigenous peoples in Alaska. Compiled from information in Champagne (1994), p. 295. Reprinted with permission of Gale Research, Inc.*

speak languages from a different linguistic family usually termed Athabascan but occasionally referred to as Na-Dene. Most Athabascan-speaking groups are seen as genetically distinct from Eskimos; however, some researchers see the two major Athabascan groups of the region (Koyukon and Gwich'in) as more closely related genetically to Alaskan Eskimos than to other Indians (Szamarthy 1984).

Eskimo subsistence patterns

Inupiat and Yup'ik subsistence patterns in the region can be categorized into four types:

1. Salmon fishing (Yup'ik groups on the lower portion of the Yukon River). Salmon represent the major focus of subsistence activities of these groups. Salmon are supplemented by seals (Phocidae), other freshwater and saltwater fish, terrestrial mammals, and migratory waterfowl.
2. Mixed hunting and fishing (Inupiat groups of the bays and lagoons, Norton Sound, and Kotzebue Sound). Groups practicing this strategy combine seals and belugas (*Delphinapterus leucas*); salmon, herring (*Clupea harengus*), and other fishery resources and caribou as subsistence mainstays.
3. Caribou hunters (Inupiat groups of interior North Alaska [Nunamiut] and the interior of the Seward Peninsula [Kauwerak]). These groups are overwhelmingly dependent on caribou (80%). Supplementary resources included Dall sheep (*Ovis dalli*), freshwater fish, hare (*Lepus* spp.), and furbearing mammals.
4. Large sea mammal hunting (Inupiat groups along the northwestern Alaskan coast and on the Bering Straits Islands, and Siberian Yup'ik groups on St. Lawrence Island). Bowhead whales (*Balaena mysticetus*), walrus, seals, and belugas were the primary focus of subsistence activities. On the mainland, supplemental food included caribou, freshwater fish, and migratory waterfowl. On the islands, supplements were freshwater and saltwater fish.

Athabascan subsistence patterns

Subsistence patterns of the Athabascan in the region are less varied than those of their Eskimo neighbors because they do not occupy coastal or insular areas, and resources within their areas tend to be less abundant. Their subsistence strategies can be categorized into three types:

1. Salmon fishing (Ingalik and Lower Koyukon groups along the lower portion of the Yukon River). Salmon are the primary focus of subsistence activity supplemented by caribou, moose (*Alces alces*), furbearers, and freshwater fish.
2. Mixed hunting and fishing (Koyukon of the central Yukon and lower Koyukuk River, Gwich'in of the central Yukon). These groups combine

salmon and freshwater fishing with hunting of caribou and moose, small
furbearers, and migratory waterfowl.
3. Caribou hunting (Koyukon of the upper Koyukuk River and Brooks
Range, Gwich'in of the Porcupine River and Brooks Range). These
groups depend on the caribou, supplemented by moose, Dall sheep, hare,
furbearers, and anadromous and freshwater fish.

Settlement patterns

Traditionally, settlement patterns varied on a continuum from sedentary permanent residence at a single location to nomadic. Settlement pattern variables
include the size of settlements and the nature of the shelters. Settlements included
permanent villages, occupied nearly year-round and containing substantial and
generally well-built structures for habitation as well as other uses (storage, cele-
bration). *Seasonal villages* were occupied for shorter periods usually associated
with the harvest and processing of resources; structures generally were of lesser
quality than those at permanent villages. *Seasonal* and/or *short-term camps* were
occupied for relatively short periods and consisted of impermanent structures
such as tents.

Eskimo settlement patterns

Three basic settlement patterns characterized Eskimo populations:

1. Medium to large permanent villages (300–1000 occupants) were found at
key locations along the northwestern Alaskan mainland and on St.
Lawrence Island. Residents pursued marine mammals from these communities. Seasonal campsites occupied by families were used for other
harvesting activities.
2. Small to medium permanent villages (50–300 occupants) with seasonal
villages and campsites were found in Kotzebue and Norton Sound regions
and along the lower Yukon River. People moved from the permanent winter village to several seasonal villages or campsites as part of an annual
round.
3. Small permanent villages (20–50 occupants) with several seasonal villages and numerous campsites were found in the interior North Alaska
and Seward Peninsula regions where Inupiat depended on caribou.

Athabascan settlement patterns

Athabascan settlement patterns also take three basic forms:

1. Small to medium permanent villages (50–200 occupants) with one or two
seasonal villages or campsites were found on the lower Yukon where
salmon was the subsistence mainstay.

2. Small permanent villages (20–50 occupants) with several seasonal villages or camps were found among the Koyukon on the central Yukon and lower Koyukuk rivers and among the Gwich'in of the Yukon Flats region.
3. Seasonal villages and campsites (less than 50 occupants) requiring relatively frequent moves in pursuit of moose or caribou were found among the upper Koyukon and the Gwich'in of the Brooks Range.

Large communities were occupied for several hundred years prior to Euro-American arrival at such locations as Tigara (Point Hope), Kingigin (Wales), and Sivukaq (Gambell). These areas are still the home of populations with high levels of subsistence dependence. The Arctic coastal plain region east of the Colville River to the Mackenzie River had much less utilization.

Linkages between subsistence activities and the cultural process were important in traditional societies. Coordination of subsistence activities was done by successful hunters and fishermen who achieved leadership standing in their communities by sharing their products. Major celebrations were characterized by distribution and consumption of subsistence products. Status and reputation were based on the abundance individuals and communities could provide on these occasions. Territorial and property rights were well-developed, reflecting a significant degree of competition for resources.

Historic and cultural change: 1850–1990

Euro-American contact in the nineteenth century brought dramatic change to Alaskan indigenous people. The introduction of diseases such as smallpox, measles, and influenza reduced populations by 50–90% (Fortuine 1989; Wolfe 1982). Other factors also reduced Native numbers. For example, the numbers of Sivukakmiut of St. Lawrence Island were reduced from 1500 to 300 from 1878 to 1880 (Bockstoce 1989). Reasons for this decline include a substantial reduction of the walrus population by Yankee whalers, and the onset of a series of stormy, cold winters making walrus hunting virtually impossible (Bockstoce 1989). Eventually, settlement patterns were dramatically altered by population reduction and interaction with the world economic system.

Many settlements disappeared as surviving populations clustered together. On St. Lawrence Island, a population previously dispersed in six to eight villages became a single village, Gambell. The interior of North Alaska was virtually depopulated as Nunamiut groups moved to the coast to replace those who had succumbed to various diseases in the 1880s and to be closer to trading opportunities.

The establishment of trading posts, such as St. Michael's in Norton Sound and Fort Yukon, led to adjustments in subsistence and settlement patterns among Native groups. For example, the village of Stebbins, founded by a group of Yup'ik from Nelson Island, relocated to be closer to trading opportunities at St.

Michael's. Annual cycles of activity had to incorporate periods for trapping, processing, and transporting hides to trading posts, which required adjustments in subsistence pursuits.

The tendency toward sedentism continued from the late nineteenth to mid-twentieth century when the last Alaskan nomadic groups settled in Lime Village and Anaktuvuk Pass. The coming of missions, the requirements of schooling, and the need for access to trade goods such as guns, ammunition, nets, traps, fuel, clothing, and food were behind the transition to permanent settlements. People still returned to their seasonal villages and campsites where resources were available, but for shorter periods.

The 1960s and 1970s brought about two developments that continue to have implications. First, improved health care reduced infant mortality while birthrates remained high, resulting in a substantial population increase in Native villages and towns. Second, small-scale motorized transportation (especially the snowmobile) replaced dogs in most villages in the 1960s. The reduction in the number of dogs reduced subsistence demand for chum salmon (*Oncorhynchus keta*) and seals that had been used for dog food. The outboard engine and skiff were also quickly adopted, and they transformed subsistence and settlement patterns by reducing travel time and greatly extending the areas that could be reached from permanent communities. Consequently, an expanding indigenous population not only maintained its subsistence activities but extended them to areas that had been unused since earlier population decimation.

Use of new technology meant that hunters needed cash to acquire and operate their equipment. With the collapse of world fur markets in the 1950s, many Inupiat, Yup'ik, and Athabascans left their villages on a seasonal or semipermanent basis to obtain the cash necessary to sustain themselves in their resident communities (Van Stone 1962, 1974). An alternative strategy to leaving local villages was made possible by the Great Society programs of the 1960s and 1970s, and the oil revenues of the late 1970s and early 1980s. Cash income became available from local construction and service expansion in the villages and towns. The North Slope borough, the local government for over 50% of the area under consideration, was formed to provide tax revenues to create local jobs and enhance living conditions for North Slope Inupiat (McBeath 1981; Kruse 1982). It has achieved those objectives, but not without such consequences as attracting a large non-Native sector, increasing familial competition for wage labor, and eroding the sense of self-sufficiency in some people (Chance 1990).

Climate and history

Evidence of climate change impacts on human populations in this region can be derived from three sources: (1) archeological evidence in conjunction with paleo-environmental data, (2) oral traditions about events occurring prior to contact, and (3) historical and ethnographic observations and findings since contact. The focus will be on evidence for the effect of climate on human populations and the natural resources on which they depend (Langdon 1984).

Analytical problems in relating climate to culture

Constructing defensible linkages between climate and human culture is problematic considering the inadequacy of critical environmental data (e.g., indicators of species and population numbers for specific periods) (Ingram et al. 1981). McGhee (1994) asserts that because of archeological data limitations, it is virtually impossible to make reliable statements connecting climate and culture. Analyses of climatic effects on societies in harsh climatic areas are probably more revealing because of more immediate vulnerability. However, linkages between climate-based impacts and postulated cultural responses are poorly understood (Ingram et al. 1981), and groups may persist in valued behaviors despite evidence that alternatives are more successful. Lines of causation that begin in climate change may initiate consequences that become more complex the more removed they are from precipitating conditions (Figure 9.2).

Archeological perspectives

Human occupation of the region is generally dated to around 13,000 B.P. The archeological record indicates that humans were primarily oriented to terrestrial hunting of relatively large herbivores during the earliest period (ca. 13,000–8,000 B.P.). With the disappearance of megafauna such as mammoth (*Mammuthus primigenius*) and giant bison (*Bison* spp.), humans appear to have shifted to more diverse food procurement strategies targeting small mammals, birds, and fish in addition to caribou and moose.

Recent archeological work in interior Alaska suggests that the period ca. 7,500–6,000 B.P. may have been drier and harsher than the present. Evidence for this proposition is based on relatively barren, sandy strata devoid of human artifacts during this period (D. R. Yesner, 1994 personal communication). Perhaps this indicates that the Altithermal period (9,000–5,000 B.P.) produced a less hospitable, warmer, and drier climate in the interior of Alaska.

There is no evidence for coastal occupations and marine resource use in northern Alaska during this period. This corresponds with a New World pattern of significant use of coastal and marine resources beginning ca. 6,000 B.P. (Yesner 1987). It cannot be assumed, however, that humans were not using coastal and marine resources, because coastal sites of that period may now be underwater. The first use of marine resources probably began ca. 6,000 B.P. on St. Lawrence Island, presumably occupied by migrants from Chukotka. There is firm evidence of marine resource use in Norton Sound at Cape Denbigh by 4,200 B.P. (Dumond 1982). The Norton Sound data suggest seasonal occupation of the coast with pop-

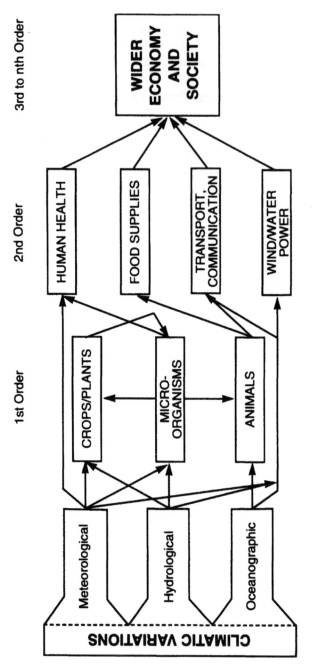

Figure 9.2. Linkages between climate-based impacts and potential sociocultural responses. Adapted from Ingram et al. (1981), figure 1.2. Reprinted with permission of Cambridge University Press.

ulations returning to interior permanent subterranean settlements during winter (Dumond 1984).

The first evidence of large marine mammal hunting is found on St. Lawrence Island ca. 3,800 B.P. Evidence for the first whaling on the north Alaska mainland dates to about 3,400 B.P. at Cape Krusenstern. This site is intriguing because it is about 1,400 years earlier than any other whaling sites on the north Alaskan mainland and its location is not within the present normal range of bowhead or gray whale (*Eschrichtius robustus*) migration. Perhaps this was a slightly warmer period than the present, allowing whales to move closer to shore in the spring. However, paleoclimatic data indicate that this was a cooler period (Chapter 2), perhaps preventing whales from reaching their customary feeding grounds in the Beaufort Sea. By 2,000 B.P., St. Lawrence Island had the technology and social organization to conduct successful large marine mammal hunts under relatively optimum conditions. This capacity does not appear on the northwest coast of Alaska until ca. 1,200 B.P.

Over the course of several millennia, a number of new technological developments enter the archeological record. Through fits and starts, extinctions and reoccupations, human populations gradually developed the capacity to utilize a wide range of resources that are available in abundance but only for short periods and in localized areas. Coastal populations developed the capacities to harvest a diversity of animals, including caribou, waterfowl, marine and anadromous fishes, and marine mammals.

This complex of capabilities is termed the Thule tradition, which appeared in western Alaska a little before 1000 A.D. Its hallmark is *flexibility* in both technology and social organization. The bearers of this tradition were apparently quite successful as they expanded eastward into the central Arctic and beyond to Hudson Bay, Labrador, and Greenland (Chapter 11). The Thule culture also expanded southward past the Kuskokwim River to Bristol Bay and the Alaska Peninsula (Dumond 1984). The Thule people were the ancestors of the more recent Inupiat populations who have exhibited the same broad range of capabilities for utilizing resources and monitoring environmental variables.

The successful expansion of the Thule across the Arctic has been linked to a changing climate. Thule expanded during a warming period that allowed walrus and bowhead whales, mainstays of the group hunting foundation of Thule, greater access to the central Arctic. This is generally thought to have occurred ca. 1000–1200 A.D. This warm period gave way to the cooler Little Ice Age period ca. 1600–1850 A.D. (Chapter 2). Dumond (1984) suggests that temperatures were as much as 5 degrees cooler on average, resulting in more sea ice and eliminating the walrus and bowhead from Arctic waters. Thule populations adapted to these conditions by developing more nomadic capabilities (as evidenced by the prominence of the ice house, or igloo); they abandoned their more permanent stone and sod subterranean lodges in order to be mobile (Burch 1979). They also developed new hunting specialties, depending on remaining local resources such as seal and caribou.

Climate change was a significant factor prodding adaptation of Thule populations to altered conditions of resource availability and temperature (McGhee 1970; Dumond 1977, 1984). Despite relatively widespread acceptance of this relationship between environmental change and population/cultural change in the High Arctic, a substantial role for social and cultural factors in mediating and directing change has also been proposed (McGhee 1994). Fitzhugh and Lamb (1985) point to cultural interactions between Inuit and Indian populations as being more influential than climate in accounting for group distribution and culture history in Labrador. Burch (1979) asserts that Inuit migration is critical in explaining new adaptive features rather than on-site positive adaptations by Inuit resident groups.

McGovern (1991, 1994) argues that cultural commitments to certain patterns of hierarchy and purity prevented the Norse from effectively adopting more appropriate material cultures and nutritional preferences from the Inuit, thus leading to their disappearance during the Little Ice Age. It is noteworthy that caribou are found in the archeological record throughout the 300-year period of decline. Caribou populations might have been expected to decline with a more severe climate. However, caribou were kept by the Norse elite after the pattern of European nobility, largely for sporting purposes even in the face of enormous nutritional deprivation. McGovern (1991) concludes, "Social stratification and a concomitant willingness and ability to impose sanctions can certainly have a positive effect on resource regulation" (p. 93).

The Norse case suggests that populations whose cultural traditions were developed in one biome may adhere to these practices despite their unsuitability for survival under different conditions. But even continuous occupation of the Arctic tundra biome and the accumulation of ecological knowledge does not guarantee that Eskimo responses to altered environmental conditions will be successful. Figure 9.3 illustrates McGovern's approach to systematically unpacking the chain of specific impacts on human subsistence likely to result from warming in the High Arctic. Research on Eskimo occupation of northeastern Greenland and subsequent disappearance from the area provides an example of this situation (Larsen 1934). Northeastern Greenland was the last area to be reached by the expanding Thule people sometime in the 1500s. Larsen's archeological reconstruction suggests that with worsening climatic conditions over the next several hundred years, they abandoned whaling of bowheads and then inexplicably moved *northward* in the eighteenth century. Larsen (1934) concludes, "Decimated in numbers, possibly by degeneration on account of their isolated life, these last dwellers in East Greenland have moved northwards in small groups. Some no doubt remained where they were and gradually died off whereas the others went on [further] to the north where their trail disappears. . ." (p. 172).

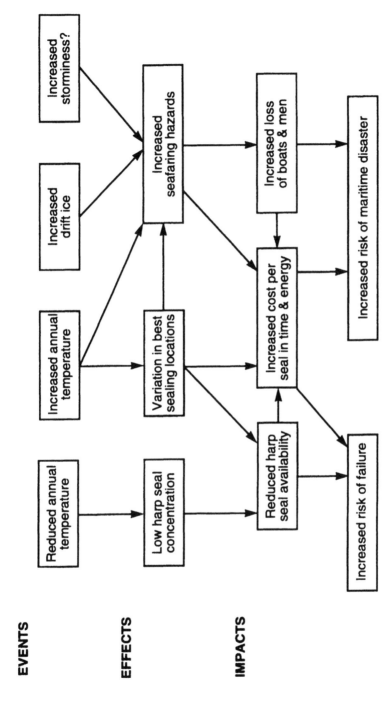

Figure 9.3. *Major linkages between climatic events, their effects on portions of the marine ecosystem, and probable impact on portions of the Norse marine hunting economy. Adapted from McGovern (1991), figure 12. Reprinted with permission of University of Wisconsin Press.*

149

Archeological and paleontological data indicate that climate alters the conditions of human existence and promotes expansion, contraction, and adaptation. However, social and cultural factors are critical in determining specific adjustments to climate change. In addition, cultural adaptations to changing climatic conditions are not necessarily automatic or successful. The Thule (and subsequently Inuit) tradition may represent the culmination of indigenous Arctic adaptation because it incorporates the technologies, knowledge, and institutions developed in previous epochs. It thus allows adjustment to a wide variety of circumstances ranging from the sedentary, large villages with complex hunting strategies capable of capturing large marine mammals, to small nomadic groups capable of sustaining themselves in winter by killing seals.

Krupnik (1988) suggests that the regular discontinuities that appear archaeologically in all but a few sites (Gambell, Wales, Point Hope) are the result of resource failure and subsequent reoccupation. He suggests that resource failures of walrus in the eastern Siberian Yup'ik communities were the result of excessive human harvesting. However, Amsden (1979) suggests that it is unrealistic to presume that human predation has been the primary factor in the periodic reductions of animal populations through time.

Oral tradition

There are oral traditions in many groups about periods of extreme deprivation and the resulting starvation or relocation of populations.

In her examination of 17 Inuit accounts of migration where causation was identified, Rowley (1985) concludes that 7 cases resulted from environmental stresses while 10 others were the result of societal pressures. In instances where Inuit populations have been forced to relocate, oral tradition often records what was perceived as the reasons for the group to move. Fienup-Riordan (1986) has recorded several tales of resource scarcity from the Yukon Delta Yup'ik. Rather than encoding detail about the nature of environmental disruption and human response, these tales address aspects of human behavior and thought that violate principles of natural order that were believed to be responsible for environmental disasters. The story "War of the Bow and Arrows" recounts the warfare arising out of failure to share during a time of resource shortage. Peace resumed once cooperation and sharing were reestablished. These accounts demonstrate that in Yup'ik thought "personal responsibility....is taken for natural disruption; nature is never held to blame" (Fienup-Riordan 1986, p. 28).

Oral traditions suggest that severe resource fluctuations have devastating effects. Flooding of the major rivers seems to have been the most disruptive environmental perturbation for Yup'ik people of the Yukon-Kuskokwim delta. Nevertheless, people do not appear to be fatalistic about such outcomes and tend to emphasize the social responsibility of the individual to share and to contribute to group welfare. Perhaps it is this recognition of the possibility of disaster, accompanied by a resolve that it can be avoided by morally correct, group-supporting action, that allows Arctic people to adapt and persist.

Ethnographic and historical examples

Historical records, first-hand observer accounts, and ethnographic research contain information on the impact of climate change on Alaskan Native cultures over the past 250 years. The interpretation of cultural responses to environmental perturbation is difficult because of the substantial reduction of the Native population due to disease and oppression. This reduction caused an altered cultural landscape in which some local populations disappeared, other remnant groups were aggregated, and virtually all suffered serious disruptions in the continuity of cultural knowledge because of the loss of elders.

A variety of observers comment on the movement of Inupiat-speaking peoples termed Malemiut from the interior and northeastern portions of the Seward Peninsula south to the coast of Norton Sound to present-day Unalakeet. A substantial part of the Inupiat-speaking population of Unalakeet in the 1970s derived from one family whose ancestral home was in the Kotzebue Sound region (Ray 1975). In Ray's view, the southward movement of these people began in the 1840s as the result of two developments. First, trading opportunities expanded due to the appearance of Euro-American goods. The Malemiut took advantage of this niche, developing into itinerant traders and acting as middlemen between the Yukon and Kuskokwim River Yup'iks and the Bering Strait Inupiat. Second, eastern Norton Sound Yup'ik suffered a drastic population decline in 1838 due to smallpox (Ray 1975).

A cool period during the second half of the nineteenth century probably caused the collapse of the western Arctic caribou herd, which withdrew from the Seward Peninsula by the 1870s and disappeared from the Kobuk River valley by the 1880s (Lewis 1993). This collapse provided a "push" factor, motivating Inupiat to leave traditional territories for unfamiliar coastal areas. Lewis concludes that this is but one instance of Eskimo movement between bioregions as a normal response to resource fluctuations resulting from climatic variation.

Burch (1979) and Amsden (1979) do not relate the decline of the western Arctic caribou herd to a cooler climate but rather to long-term population fluctuation. Nevertheless, both authors believe this decline to be the major factor in the disappearance of the Nunamiut (interior North Alaskan Inupiat) from much of their territory in the Brooks Range. Amsden (1979) identifies population emigration east to the Mackenzie Delta as one of the responses to this circumstance. Those Nunamiut who did remain occupied only one-third of the previous territory at a much lower density and shifted to the east, where they had opportunities to harvest from the Porcupine caribou herd (Gunn 1994).

Amsden's (1979) analysis of "hard times" (the down period of long-term cyclical fluctuations in resource populations) is useful for illustrating the potential impacts of a warmer climate. He notes that the Nunamiut who survived and adjusted in the Brooks Range through the decline of the caribou did so by territorial adjustment and shifts to exploitation of Dall sheep and freshwater fish. However, after the rebound of the caribou herd (ca. 1910), the Nunamiut reoccupied the area and by the 1940s rebuilt an adaptation very similar to that of traditional times, despite a much altered social context (Amsden 1979). This adjustment may represent the replication of cultural templates continuing among the elder Nunamiut, or it may mean that there are limited subsistence-based ways to survive in the Brooks Range.

Subsistence strategies and harvests of Kivalina Inupiat provide evidence of the extraordinary range of harvest levels associated with indigenous adaptations. In four comparative years (1964–65 and 1983–84), *anomalies* (harvests far exceeding those of other years) occurred for one resource *in every year*. Burch (1985) concludes that anomalies are part of the *normal* cycle of Kivalina hunters. Anomalies were found at both the high and low ends of harvest levels and ranged across key resources (bowhead whale, walrus, caribou, fish, and seals). Despite an expectation of variation, two hunting and fishing periods are perceived as crucial in Kivalina: early summer when dried fish and seal are produced and early fall when salmon are taken. If these subsistence foundations fail, people consider themselves in a crisis regardless of how much other food is available (Burch 1985). Perhaps this definition of crisis serves as a signal to refocus efforts toward other resources.

Summary

Indigenous populations in the North American Arctic and subarctic have developed significant flexibility in their adaptations to cope with fluctuations in resource availability. However, major sustained resource disruptions have caused catastrophic reduction and even disappearance of groups. Although most groups have shown their resilience and ability to survive in the face of natural hardships and shortages, they may not be so well-equipped to cope with increasing immersion in the consumer culture of the information society. Cultural patterns persist only if they are accepted as meaningful and fulfilling by each generation. There must also be an opportunity for associated activities and values to be expressed. The combination of alternative cultural lifestyles and altered subsistence opportunities resulting from a warmer climate may pose the greatest threat of all to the continuity of indigenous cultures in northern North America.

References

Amsden, C. W. 1979. Hard times: A case study from northern Alaska and implications for Arctic prehistory. In *Thule Eskimo culture: An anthropological retrospective*, edited by A. McCartney, 395–410. Paper 88. Ottawa: Archaeological Survey of Canada.

Bockstoce, J. R. 1989. *Whales, ice and men: A history of whaling in the western Arctic.* Seattle: University of Washington Press.

Burch, E. S. 1979. The Thule-historic Eskimo transition on the west coast of Hudson Bay. In *Thule Eskimo culture: An anthropological retrospective*, edited by A. McCartney, 189–211. Paper 88. Ottawa: Archaeological Survey of Canada.

Burch, E. S. 1985. *Subsistence production in Kivalina, Alaska: A twenty-year perspective.* Division of Subsistence Technical Paper 128. Juneau: Alaska Department of Fish and Game.

Champagne, D., ed. 1994. *Native North American almanac.* Detroit, MI: Gale Research, Inc.

Chance, N. A. 1990. *The Inupiat and Arctic Alaska: An ethnography of development.* Fort Worth, TX: Holt, Rinehart, and Winston.

Dumond, D. E. 1977. *The Eskimos and Aleuts.* London: Thames and Hudson.

Dumond, D. E. 1982. Trends and traditions in Alaska prehistory: The place of Norton culture. *Arctic Anthropology* 19:39–51.

Dumond, D. E. 1984. Prehistory: Summary. In *Handbook of North American Indians, Vol. 5—Arctic*, edited by D. Damas, 72–79. Washington, DC: Smithsonian Institution Press.

Fienup-Riordan, A. 1986. *When our bad season comes: A cultural account of subsistence harvesting and harvest disruption on the Yukon Delta.* Aurora Monograph Series 1. Anchorage: Alaska Anthropological Association.

Fitzhugh, W., and H. Lamb. 1985. Vegetation history and culture change in Labrador prehistory. *Arctic and Alpine Research* 17:357–370.

Fortuine, R. 1989. *Chills and fevers: Health and disease in the early history of Alaska.* Fairbanks: University of Alaska Press.

Gunn, J. D. 1994. Global climate and regional biocultural diversity. In *Historical ecology: Cultural knowledge and changing landscapes*, edited by C. L. Crumley, 67–97. Santa Fe, NM: School of American Research Press.

Ingram, M., G. Farmer, and T. M. L. Wigley. 1981. Past climates and their impact on man: A review. In *Climate and history: Studies in past climates and their impact on man*, edited by T. Wigley, M. J. Ingram, and G. Farmer, 3–50. New York: Cambridge University Press.

Krupnik, I. I. 1988. Asiatic Eskimos and marine resources: A case of ecological pulsations or equilibrium? *Arctic Anthropology* 25:94–106.

Kruse, J. 1982. Energy development on Alaska's north slope: Effects on the Inupiat population. *Human Organization* 41:97–106.

Langdon, S. J. 1984. *Alaska Native subsistence: Current regulatory regimes and issues.* Alaska Native Review Commission, vol. XIX. Anchorage: Alaska Native Review Commission.

Larsen, H. 1934. Dodemandsbugten—An Eskimo settlement on Clavering Island. In *Meddelelser om Gronland*, Vol. 102(1). Copenhagen: C. A. Reitzels Forlag.

Lewis, M. A. 1993, April 8–10. Climate correlates of the 19th century Inupiat population movement from Kotzebue Sound to Norton Sound as reconstructed from tree-rings. Paper presented at the Alaska Anthropology Association 20th Annual Conference, Anchorage.

McBeath, G. 1981. *North Slope Borough government and policymaking*. Fairbanks: Institute for Social and Economic Research, University of Alaska.

McGhee, R. 1970. Speculations on climatic change and Thule culture development. *Folk* 11/12:173–184.

McGhee, R. 1994, April 10–14. Effects of past climate change on peoples of the North. Paper presented at the Mid Study Workshop of the Mackenzie Basin Impact Study, Yellowknife, NWT.

McGovern, T. H. 1991. Climate, correlation, and causation in Norse Greenland. *Arctic Anthropology* 28:77–100.

McGovern, T. H. 1994. Management for extinction in Norse Greenland. In *Historical ecology: Cultural knowledge and changing landscapes*, edited by C. L. Crumley, 127–154. Santa Fe, NM: School of American Research Press.

Ray, D. J. 1975. *The Eskimos of Bering Strait, 1650–1898*. Seattle: University of Washington Press.

Rowley, S. 1985. Population movements in the Canadian Arctic. *Etudes Inuit Studies* 9:3–21.

Szamarthy, E. J. 1984. Human biology of the Arctic. In *Handbook of North American Indians, Vol. 5—Arctic*, edited by D. Damas, 64–71. Washington, DC: Smithsonian Institution Press.

Wolfe, R. 1982. Alaska's great sickness, 1900: An epidemic of measles and influenza in a virgin soil population. *Proceedings of the American Philosophical Society* 126:1340–55.

Van Stone, J. W. 1962. *Point Hope: An Eskimo village in transition*. Seattle: University of Washington Press.

Van Stone, J. W. 1974. *Athabascan adaptations: Hunters and fishermen of the subarctic forests*. Chicago: Aldine Press.

Yesner, D. R. 1987. Life in the "Garden of Eden:" Causes and consequences of the adoption of marine diets by human societies. In *Food and evolution: Toward a theory of human food habits*, edited by M. Harris and E. Ross, 285–310. Philadelphia: Temple University Press.

10

Resource use in rural Alaskan communities

Donald Callaway

Indigenous people have subsisted on natural resources in the Far North for thousands of years. Although technology has changed and cash now plays an important role, this dependence on resources continues today, not only in Alaska Native villages but in many non-Native rural households as well.[1]

Where Alaskans live

About 70% of Alaska's population of 550,000 reside in the metropolitan areas of Anchorage, Fairbanks, and Juneau. Only 17 other municipalities have populations greater than 2,500. In total, there are 23 hub or minor hub (subregional) communities with a combined population of about 80,000. These communities vary greatly in their ethnic characteristics, from being predominantly Native (e.g., Kotzebue) to predominantly non-Native (Kenai). If residents of the hub communities are added to the metropolitan population, about 84,000 people remain. These Alaskans live in the rest of the state, including in 210 Native villages, most of which are inaccessible by road.

Wildlife harvest characteristics in rural Alaska

Dietary importance

People in most regions of Alaska harvest more meat and fish per capita than is purchased by the average United States citizen (Figure 10.1). The metropolitan

---

[1] In modern regulatory terms, subsistence is defined as a customary and traditional use of naturally occurring wild resources for food, clothing, art, crafts, sharing, and customary trade. The Alaskan and federal governments have adopted eight factors or criteria that help identify customary and traditional subsistence users: (1) long-term, consistent pattern of use (including duration and consistency of use), (2) use pattern recurring in specific seasons, (3) methods and means of harvest including the composition of hunting parties, (4) harvest near or reasonably accessible, (5) handling, preparing, preserving, and storing of resources, (6) handing down of knowledge, (7) distribution or sharing of resources, and (8) reliance on a wide diversity of wild resources. Each of these criteria is divided into historic and contemporary use.

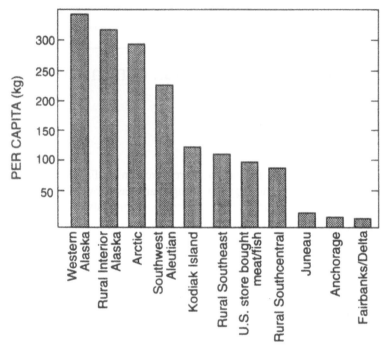

Figure 10.1. Per capita wild resource harvest in selected regions of Alaska. Data are from Wolfe and Bosworth (1994).

areas of the state (Anchorage, Juneau, Fairbanks) harvest small per capita amounts of wildlife resources in addition to purchasing meat, fish, and poultry in quantities that are typical for the rest of the United States. The reverse is true for most of rural Alaska, especially for small rural communities, where modest amounts of purchased goods supplement a diet composed mostly of wildlife resources.

Most households in rural areas participate in the harvest of wild foods (Figure 10.2). In most areas, over half of the households harvest game (primarily large land mammals), and a higher proportion harvest fish. An even higher proportion of households *uses* resources rather than (or in addition to) *harvesting* them. Many households that are incapable of harvesting resources (e.g., the elderly) still use wild foods by virtue of sharing with other households.

Approximately 20 million kilograms of wild foods are harvested annually by residents of rural areas. Annual harvest is 170 kilograms per person, compared to 10 kilograms per person for urban areas (Wolfe and Bosworth 1994). Fish provide the largest contribution to subsistence diets throughout rural Alaska. However, rural communities in the northern part of the state depend more on marine mammals than on fish, and many interior communities depend more on game. Communities in western Alaska depend heavily on fish.

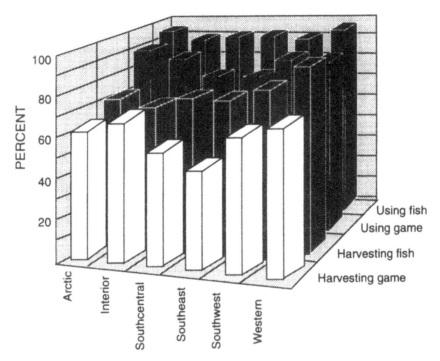

Figure 10.2. Rural households participating in subsistence in different regions of Alaska. Data are from Wolfe and Bosworth (1994).

Economic importance

The relationship of household income to the harvest of wildlife resources for Alaskan households is complex. On one hand, full-time wage employment often provides the income necessary to purchase the technology (guns, boats, motors, nets, etc.) needed to engage in harvesting activities. On the other hand, full-time employment often restricts the time necessary to engage in harvest activities. Technology allows quick access to resources, and full-time wage earners often share their harvest technology with other household members or relatives.

Traditional technologies are insufficient for current harvesting techniques, so households require a significant income to participate in subsistence activities. Obtaining this income through wage employment, commercial fishing, or government transfers is a considerable challenge in rural communities that contain few jobs. The replacement cost of harvested wildlife resources would be enormous, especially considering distance from markets, cost of transportation, and the lack of retail sectors in most small rural communities. Assuming a replacement expense of $6–11 per kilogram, the annual replacement cost of the wild food harvest in rural Alaska is estimated at $131–218 million.

The per capita cost of replacing subsistence foods ($3,063) would be 49% of per capita mean incomes for rural interior Native families (Wolfe and Bosworth 1994). In other regions, theoretical replacement costs vary from 25 to 50% of total family income. The cost of purchasing harvest technology is not well-documented for Alaska. In some cases (e.g., the community of Gambell on St. Lawrence Island), nearly half of an average household's available income may go toward the purchase of harvest technology and fuel (Robbins and Little 1984).

Resource diversity

Resource availability and use vary greatly on a regional basis, as seen in Table 10.1. In general, the greater the diversity of food types used, the more likely a community is to exhibit a customary and traditional (i.e., consistent with the eight cited criteria) profile of resource use. On this diversity index, five species is a useful criterion for distinguishing between traditional and nontraditional resource patterns for communities of interior Alaska. Eight species is more appropriate for coastal communities with access to more resources.

A community such as Ivanof Bay—a small, relatively homogeneous community populated by long-term residents with access to a variety of land and marine resources—lies at one end of a continuum. Manokotak and Port Graham share many of these characteristics, although Manokotak has access to marine resources because of travel (and trade) through riverine and estuarine environments. Point Lay, which harvests less than half as many species as Ivanof Bay, has higher per capita and household harvests. Communities that harvest the lowest variety of resources generally have an average household harvest of less than 230 kilograms. The eight factors adopted in federal regulations to define "customary and traditional" subsistence practices include a measure of the diversity of species harvested by a community, specifically a use pattern that includes reliance for subsistence purposes on a diversity of the fish and game resources.

Table 10.1. Harvest levels and species diversity for selected Alaskan communities.

	Number of species used by >50% of households	Household total (kg)	Per capita (kg)
Ivanof Bay	32	759	207
Manokotak	27	911	174
Port Graham	18	296	103
Point Lay	13	1134	404
Tetlin	10	388	97
Kodiak City	9	224	67
Slana	6	257	113
Tok	4	200	67
Glenallen	3	124	45

There are different ways to define resource diversity. A cumulative index simply summarizes the number of species used by at least one household in the community and provides an overview of the total number of resources used by the community (Wolfe 1992). For example, Tatitlek (a small Native community on Prince William Sound) uses 52 different species of wild foods. One drawback to this measure is that it does not express any variability in the distribution of resource users. It is possible that only a few households in Anchorage could harvest a sufficient variety of resources to give it the same score as Tatitlek. This would result in the misleading conclusion that the two communities are similar in the variety of resources used.

Wolfe (1992) standardizes this variability with another measure: the number of different species of wild resources *used by at least half of a community's households*. In this case, Tatitlek's *cumulative* resource index of 52 different types drops to a *standardized* score of 19 different resource types. Kotzebue, a modest-sized community of 3,500 people with a cumulative index of 52 total resources, drops to *standardized* score of 6. The diversity of use by a majority of households in Anchorage would probably be one or two. Data on standardized diversity of use for five communities in the Upper Tanana River valley[2] are summarized in Table 10.2.

The majority of households in Tok utilize the fewest resources, and harvests are clearly dominated by big game with no small game, salmon (*Oncorhynchus* spp.), or plants and berries. Segments of Tok's population have a sports-hunting orientation and depends on only a few types of resources. The other communities in the Upper Tanana region use a broad range of resources with the exception of salmon, which is not locally available in many places. Thus with the exception of Tok, all communities depend heavily on a diverse harvest of fish, big game, birds, and plants and berries.

Table 10.2. Standardized diversity of wild food use (number of species used by >50% of households) is summarized by resource category for five Upper Tanana communities.

	Tok	Northway	Tanacross	Tetlin	Dot Lake
Big game	2	2	2	1	2
Birds/eggs	1	2	1	1	1
Non-salmon fish	1	3	3	3	4
Plants and berries	0	2	2	3	3
Salmon	0	0	1	0	1
Small game/furbearers	0	1	1	2	1
Total	4	10	10	10	12

[2] Tok (367 households) is a much larger community than other communities in the Upper Tanana region. Tok also has the lowest proportion of Native residents and persons born locally. Dot Lake (20 households) is the smallest community, formed by a group of Native families descended from Athabascan bands, with immigrants from outside the state residing in households along the highway. Northway, Tanacross, and Tetlin are small predominantly Native communities in which most people are life-long residents.

Some areas have a greater diversity of harvestable resources than others. For example, coastal communities in Alaska have greater resource diversity than inland communities because of the availability of marine fish, invertebrates, and mammals. Thus a community that utilizes 10 out of 20 available resources (50%) can be viewed as having a proportionally greater diversity of uses than a community that uses 15 out of 45 available resources (33%). Both absolute and relative measures of diversity need to be considered in assessments of subsistence harvest information.

The cultural importance of subsistence in rural communities

Subsistence resources do more than supply nutrition to small rural (predominantly Native) communities in Alaska. Participation in family and community subsistence activities—whether it be clamming, processing fish, or seal hunting—provides strong memories and basic values in an individual's life. These activities define and establish a sense of family and community. It is also a process through which individuals learn to identify, harvest, and process a wide variety of resources. Food preferences are strongly expressed features of all cultures; the preparation and special taste of foods encountered by children as they grow up stays with them forever.

The distribution of food resources establishes and promotes the most basic ethical values in a culture: generosity, respect for the knowledge of elders, self-esteem for the successful harvest, and family and public appreciation for distribution of the harvester. No other set of activities provides a similar moral foundation for continuity between generations. Kruse et al. (1983) cite several reasons why subsistence is important to the Inupiat way of life:

- Inupiat youth must learn about their subsistence livelihood on a continual basis. This allows them to not only preserve important aspects of Inupiat culture, but to survive in a subsistence-based economy during anticipated economic downturns.
- Changes in natural resources due to inappropriate human intervention (such as pollution or overharvesting) violates Inupuiat respect for nature.
- Interruption of subsistence pursuits influences the ability of Inupiat to cooperate and share in traditional ways.
- Loss of Native foods is detrimental because: (1) Inupiat prefer the taste of Native foods, (2) sharing Native foods provides reciprocity and support within and between generations, and (3) Native foods are more nutritious.
- Many Inupiat feel subsistence products are essential for self-reliance and think that total dependence on "Anglo food" is unwise.
- The ability to hunt, share, and consume Native foods is an affirmation of cultural identity and well-being.

Subsistence and social organization

The social organization of Native Alaskan groups (and many non-Native households) encompasses households, families, and wider networks of relatives and friends. These various types of networks have overlapping memberships and have linkages related to the extraction, distribution, and consumption of natural resources.

There is generally a much greater level of interdependence among Alaska Native households than among urban households. There is also a sharp contrast between Native and non-Native (including rural) cultures: Native people expect and need support from extended-family members on a day-to-day basis. Native cultures have strong expectations about sharing; households are not necessarily viewed as independent economic units. Giving by successful community members (e.g., hunters) is regarded as an end in itself, although community status and esteem accrue to generous givers.

The changes and stresses encountered by independent nuclear households in the United States are typically addressed by diverting a portion of a household's cash to purchase services. There is also an expectation that the economy will provide a steady income for at least one adult household member and reasonable opportunity for at least part-time employment for a spouse. Neither of these expectations exists for small rural communities in Alaska.

Subsistence, sharing, and cultural identity

In reference to their work with Native Alaskan households on St. Lawrence Island, Robbins and Little (1984) note that "It is crucial to realize that each household is severely limited in its ability to marshal human and natural resources, that are indispensable for existence, without the cooperation of other households" (p. 101). The necessity of sharing between households in aboriginal times was an adaptive mechanism for dealing with variability of access to subsistence resources. The current sporadic wage economy does not provide sufficient income for most households to act as independent units even if they wanted to. Considerable labor is needed for subsistence activities, and most households have low cash incomes, so there is still motivation for cooperative activities. Subsistence activities create networks that draw households, families, and villages together (Robbins and Little 1984). The importance of sharing is further emphasized by Kruse et al. (1981):

> Only one aspect of community life that North Slope Eskimos recalled as having been among the best in 1970 remained among the best in 1977: villagers still helped each other and shared with each other in 1977. (p. 82)

The extent of sharing networks among Native households

The importance of inter-household sharing networks is described by Robbins and Little (1984, p. 112). They describe a household that consists of a man, who is a wage-earner several months out of the year; his wife; and their three teenage children. The man's wage job prevents him from engaging in extensive hunting and fishing, although he hunts on the weekends alone, with his son, or with his brothers. He participates in about half of the walrus (*Odobenus rosmarus*) and whale hunts. His brothers and other male kin, who are without wage jobs, hunt at every opportunity. His wife, who is responsible for preparing food and distributing subsistence products, distributes about 50% of the subsistence products that are produced. The proportion dispensed is average for a good hunter; the best hunters share 70% of their take, and excellent unmarried hunters may redistribute 90% of their harvest. The majority of the diet of St. Lawrence households consists of subsistence products. The first people to receive these products are elders, widows, people without hunters in their households, the ill, and the infirm. The recipients are typically relatives, but kin affiliation is not a necessary condition for receiving subsistence products. The wife in the example ensured that six households received a steady supply of subsistence resources. These were headed by elderly women and an elderly widower. Various amounts of goods were gifted to 29 households in Gambell and 23 in Savoonga, as well as to extended-family members who now live in Fairbanks (Alaska), Nome (Alaska), Sitka (Alaska), California, and Oregon.

These gifting and reciprocal relations span lifetimes and are characteristic of small communities originally based on hunting and gathering economies. Gifting during one stage of one's life (young adult to middle-age) is complemented by receiving during other stages (childhood, adolescence, and old age). The profound cultural value of sharing is described extensively in the ethnographic literature (Nelson 1969, 1982; Burch 1975; Galginaitis 1984; Chance 1990).

The complex relationship among cultural values, subsistence pursuits, and the risks posed by changing resource conditions (as a result of climate change or other factors) is illustrated in testimony of North Slope Borough community member Horace Ahsogeak:

> And I cannot fulfill the role of an Inupiat hunter that I have been taught to do....that I must always share what I hunt with poor people who cannot hunt. Already the hunting is getting so difficult that it is hard for me to continue the sharing I want and need to do to be a true Inupiat hunter. (Kruse et al. 1983, p. 234)

The sharing of wildlife resources

Inupiat Ilitqusiat (contained within the Nana Regional Strategy, Maniilaq Association, 1982) presents a distillation of Inupiat traditional values. "Every Inupiaq is responsible to all other Inupiat for the survival of our cultural spirit, and the values and traditions through which it survives. Through our extended family, we retain, teach, and live our Inupiaq way" (p. 86). The values enumerated in *Inupiat Ilitqusiat* include: knowledge of language, sharing, respect of others, cooperation, respect for elders, love for children, hard work, knowledge of family tree, avoidance of conflict, respect for nature, spirituality, humor, family roles, hunter success, domestic skills, humility, and responsibility to tribe. Several of these values directly or indirectly speak to the key value of sharing in Alaskan Native (and rural) culture.

Sharing in a harsh and uncertain environment is a practical cultural characteristic. The distribution of successful harvests is critical to the survival of the community, especially in times of need. Given the uncertain nature of resources in an Arctic environment, no one hunter or family is always successful. Institutionalized sharing provides a safety net during times of uncertain resource availability. Furthermore, sharing is an integral part of individuals' perspectives within the context of the overall family cycle of the community. People are regarded not as a burden (barring health problems) but as a soon-to-be or former contributor. Sharing and giving are lifetime expectations, and there are profound social reinforcements for these behaviors.

Table 10.3 summarizes the harvest, gifting, and reception of wildlife for communities in the Upper Tanana region. Tetlin has the highest percentage (87%) of households that gifted wildlife resources, and one of the lowest proportions of households that did not receive harvested resources. Thus Tetlin represents one of the more traditional communities in this region. This community contrasts with Tok, in which low harvesters (defined subjectively by community residents as those individuals who are not known to harvest large amounts of wild foods; this contrasts with high harvesters, who have a reputation for procuring large amounts of wild foods) constitute about two-thirds of its population. The low harvesters in Tok are predominantly receivers of wildlife harvests and give fewer resources than they harvest or receive. Non-Natives generally give fewer resources than Natives.

Nonetheless, high harvesting households in Tok, including non-Native households, have much in common, including the sharing of resources; there are both Native and non-Native high harvesting households in all of the Upper Tanana communities. The clear division in the region (using both harvest practices and social criteria) is between: (1) Tok low harvesters, (2) non-Native households in Tanacross, Dot Lake, and Northway, and (3) Tok high harvesters and Native households. As the age of the head of a household increases, the number of resources received by that household increases, especially for Native households in which elders may have difficulty obtaining their own food.

Climate change, cultural change, and access to wildlife resources in Alaska

The potential effects of climate change on northern ecosystems are contained in the first two sections of this book. Of special interest are possible effects on ungu-

Table 10.3. Sharing of harvested resources (mean value) in five Upper Tanana communities.

	Tok low harvesters[a]	Combined communities[b]		Tok high harvesters		
		Non-Native	Native	Non-Native	Native	Tetlin
Number of resources harvested	4.2	8.8	7.9	11.6	9.9	8.1
Number of resources received	2.7	2.5	4.3	2.2	2.9	3.7
Number of resources given	0.2	2.1	3.2	2.2	2.4	2.6
Household harvest (kg)	91	240	389	426	451	483
Per capita harvest (kg)	33	67	121	134	127	119
Households giving (%)	13	38	66	63	62	87
Households receiving no resources (%)	22	20	4	17	21	7

[a] Tok low harvesters are all non-Native.
[b] Combined communities of Dot Lake, Northway, and Tanacross

lates (Chapter 6), marine mammals (Chapter 7), and anadromous fish (Chapter 8)—all major subsistence resources. Natural fluctuations in these resources have been documented historically. In addition, contemporary management regimes, including the introduction of hatcheries, the imposition of seasons and bag limits, and a bureaucratic administrative structure, have already altered traditional responses to natural fluctuations. Two questions arise in dealing with the consequences of climate change on subsistence resources. First, can Western science predict the outcomes of climate change? Second, can current management systems respond to changes in natural resource populations?

I am reminded of the biologist who, when discussing the complex food web associated with the Bering Sea, stated "we don't even know what to measure (i.e., variables), and even if we did (know what to measure) we would not know how to measure them." This comment at the Department of Interior, Minerals Management Service's Information Transfer Meeting in 1991 was made with respect to the dramatic decrease in Steller sea lion (*Eumetopias jubata*) and red-legged kittiwake (*Rissa brevirostris*) populations in the Bering Sea. Something was happening to these populations, as evidenced by reports from Native hunters who depend on these sea mammal populations and from agency biologists who were monitoring the species.

The presumption is that something is affecting both populations, but that the exact cause is unknown. There is some evidence that the sea lion population is nutritionally stressed. Some contend that commercial fisheries have depleted prey stock as evidenced by a decline in populations of pollock (*Pollachius chalcogrammus*) and other small fish. However, is this decrease due to commercial fishing or changes in the productivity of the fishery linked to changing thermal conditions in the water? The changes may simply be due to some long-term natural cycle. Are the pollock responding to other changes in their food sources that may not be related to fishing or thermal conditions? Is the declining sea lion population related to increased predation by sharks and other predators? Does a combination of these factors explain the decreases?

Not knowing what to measure or how to measure it makes an empirical ecosystem approach difficult. Another example is the recent decline in Yukon-Kuskokwim chum salmon (*Oncorhyncus keta*) runs, which are used by subsistence and commercial fishermen. Biologists failed to predict the poor return, and upriver subsistence and commercial fishermen were forced to close their fisheries to allow for a sufficient escapement of fish. Inability to predict the run was blamed on an excessive catch by commercial fisherman, on an unexpected decrease in productivity in the feeding grounds, and on an artificially low count of salmon caused by misuse of side-scan sonar techniques. The only clear culprit established so far has been the faulty set-up of the sonar. Many other examples could be cited to illustrate that we do not have enough information to predict variations in animal populations under normal conditions, much less to accurately predict how climate change will affect them.

Subsistence hunters, who have keen observational skills along with a rich oral history tradition, are superb natural historians (Merculieff 1990 [Native viewpoint]; Morrow and Hensel 1992 [non-Native viewpoint]). Traditional eco-logical knowledge (Chapter 15) may improve our assessments of changing environmental conditions. Unfortunately, traditional knowledge is usually ignored by resource managers whose commitment to population models limits their willing-ness to consider other forms of information and insight.

An analysis of transcripts of a Kuskokwim River Salmon Management Working Group are particularly instructive. Non-Native managers, who have decision-making power, solicit testimony from Natives but tend to ignore it because they have a different cultural view of the world. A non-Western view of the natural world is illustrated by an emphasis on the sentience of animal species:

> While both Yupiit and Kass'at fear impending ecological disaster, their causal expla-nations differ. For Yupiit, changing weather patterns, a decrease in the frequency of supernatural encounters, and reduced fish and game stocks are all considered to be effects of human failure to respect nature. Non-Yup'ik scientists, although they may debate the policy applications of the 'conservation' concept, collectively assume that it is destruction of habitat and over-hunting which endanger biological populations. While Yupiit see animals as returning by will, biologists see them as self-renewable only through physiological processes. Both see habitat destruction as a problem, but the Yupiit abhor it as symptomatic of human disregard for animals, which prompts disregard from animals in turn. (Morrow and Hensel 1992, p. 42)

Although Western wildlife managers may speak of declining populations (e.g., fish) in their public discussions, Yup'ik participants are careful not to talk about insufficient numbers of fish:

> The idea that one's speech can 'make things happen' is well-attested in Yup'ik soci-ety. By speaking of the fish positively, Yupiit hope to assure their continuance and to prevent realization of the negative prognoses of non-Native managers. To the non-Native, however, this seems like a predictable response from people who have a vested interest in catching fish. (Morrow and Hensel 1992, p. 43)

Even circumstances in which there appears to be mutual agreement between Native and Western biologists can be misleading. For example, both agree that it is better not to band birds, collar bears, or be intrusive with respect to animal behavior:

> Many elders, in particular, regard techniques such as capturing and tagging or writing on eggs as intrusive and believe that the researchers hamper the birds reproductive success. Because disturbance of birds during the nesting season is problematic for biologists as well, it is easy for them to assume a similar basis for this complaint. In fact, however, it is not reproductive success per se which is the issue; rather, it is the

response of sentient beings who, being affronted, make themselves unavailable to humans. (Morrow and Hensel 1992, p. 43)

Summary

Climate change may result in significant alteration of northern ecosystems during the next century. Future changes in wildlife and fish populations may be viewed as a physiologically based response by some observers, and as a sentient response by others. In any case, we will need all available information—Western scientific data and traditional knowledge—to monitor and conserve sustainable resources for future generations.

References

Burch, E. S. 1975. *Eskimo kinsmen: Changing family relationships in northwest Alaska*. San Francisco: West Publishing.

Chance, N. A. 1990. *The Inupiat and Arctic Alaska*. New York: Holt, Rinehart, and Winston.

Galginaitis, M. 1984. *Ethnographic study and monitoring methodology of contemporary economic growth, socio-cultural change, and community development in Nuiqsut, Alaska*. Alaska OCS Social and Economic Studies Program Technical Report 96. Anchorage: Alaska OCS Region, Minerals Management Service.

Kruse, J. A., J. Kleinfeld, and R. Travis. 1981. *Energy development and the North Slope Inupiat: Quantitative analysis of social and economic change*. Man in the Arctic Program Monograph 1, Anchorage: Institute of Social and Economic Research, University of Alaska.

Kruse, J. A., M. Baring-Gould, and W. Schneider. 1983. *A description of the socioeconomics of the North Slope of the North Slope Borough*. Alaska OCS Social and Economic Studies Program Technical Report 85. Anchorage: Minerals Management Service.

Merculieff, L. 1990, May 30. Western society's linear systems and aboriginal cultures: The need for two-way exchanges for the sake of survival. Paper presented at the Sixth International Conference on Hunting and Gathering Societies, Fairbanks. Department of Commerce and Economic Development, State of Alaska.

Morrow, P., and C. Hensel. 1992. Hidden dissension: minority-majority relationships and the use of contested terminology. *Arctic Anthropology* 29:38–53.

Nelson, R. K. 1969. *Hunters of the Northern ice*. Chicago: University of Chicago Press.

Nelson, R. K. 1982. Harvest of the sea: Coastal subsistence in modern Wainwright. Report for the North Slope Borough Coastal Management Program. Barrow, AK: North Slope Borough.

Robbins, L. A., and R. L. Little. 1984. *Effects of renewable resource harvest disruptions of socioeconomic and sociocultural systems: St. Lawrence Island.* Alaska OCS Social and Economic Studies Program Technical Report 89. Anchorage: Minerals Management Service.

Wolfe, R. J. 1992. *Resource diversity as a characteristic of subsistence use.* Unnumbered report. Juneau: Division of Subsistence, Alaska Department of Fish and Game.

Wolfe, R. J., and R. G. Bosworth. 1994. *Subsistence in Alaska: 1994 Update.* Juneau: Division of Subsistence, Alaska Department of Fish and Game.

11

Warming the Arctic: Environmentalism and Canadian Inuit

George W. Wenzel

An environmental warming, albeit one more ecosocial than biophysical, is already under way from the perspective of Canadian Inuit.[1] While climate change may pose a future threat to natural resources and human populations, there is already considerable stress on northern human-environmental relations. The rapidly expanding gap between Inuit consumptive use of northern renewable resources and the environmental concerns of a *Qallunaat* public have significantly altered the resources and social fabric of the Canadian Arctic (Figure 11.1) (*Qallunaat* is the Inuktitut word for all non-Inuit. In recent years, it has become a descriptive word for Euro-American visitors to the North. It is used here to mean all non-Inuit, or more generally "Southerner.")

The last decade has made it plain to Inuit that their relationship to the Arctic environment, which they understand to be culturally traditional, is increasingly being impacted by the perceptions and actions of a vocal and politically adept group of "eco-constructivists." These advocacy groups understand the Far North's ecology and people not through the realities of the biophysical and social sciences, but through a narrow belief system of how humans should interact with nature (Gross and Levitt 1994). Although these groups are ideologically distinct from the mainstream of North American and European environmentalism, the environmentalist mainstream has stood silently by in discussions of many northern issues.

---

[1]The term *Inuit* properly refers to speakers of Inuktitut, or Inuit-Inupiaq (see Woodbury 1985), dialect of the Eskaleutian language family. In a geographical and cultural area sense, Inuit includes within its sweep all Eskimos living from Northwest Alaska to Greenland. It also refers here to the Eskimos of Canada, where Inuit is recognized as an official cultural and ethnic designation. It should also be noted that Inuit is plural nominative and possessive in form; the singular nominative is Inuk.

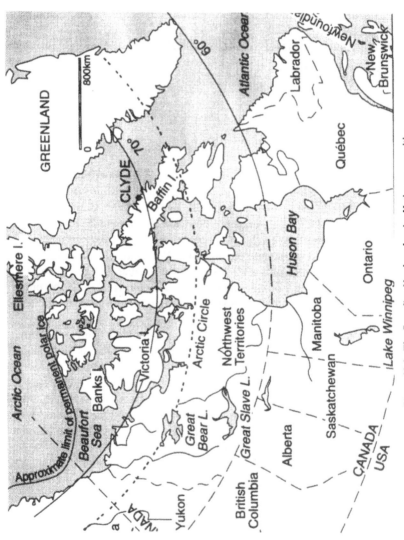

Figure 11.1. The Canadian North and major Native communities.

Global warming, science, and the Arctic

Atmospheric data suggest that the world's climate may experience a significant and possibly threatening greenhouse warming during the next century (Chapter 2). If the biophysical changes accompanying climatic warming are of the magnitude predicted (Chapters 4–8), there will be many potential cultural, demographic, and ecological effects on the Far North's indigenous peoples. The effects of past warming events on the North and its original inhabitants strongly suggest that a new northern biophysical environment will also have broad ecological impacts on Arctic marine and terrestrial faunas (Vibe 1967) and, therefore, on Inuit.

Studies of the impact of climate on northern indigenous people are rare. There are no clear cultural analogues with a content as detailed as the record of past tree line movements or postglacial marine submergences. While the transition from Dorset Paleoeskimo to Thule culture bears witness to the effects of previous Arctic warming on humans, modern Inuit are, to most observers, only pale imitations of their prehistoric ancestors.

To most people within and outside northern science, there is a perceived break between the traditional past of Inuit and their cultural present. Although it may appear that the ecological state that existed for prehistoric Thule Inuit whalers has disappeared, in reality the relationship of Canadian Inuit to wildlife is still central to modern Inuit culture and economic adaptations (Wenzel 1991; Nuttall 1992). While specific relationships with wildlife have changed, the cultural and ecological centrality of wildlife to Inuit has not.

The northern environment is already undergoing changes that have profound ecological, cultural, and economic importance for Inuit. While some of these near-term effects relate to biophysical phenomena, the major impacts on Inuit are the result of another kind of warming trend: the changing social relationship between Inuit and "the South."

Inuit culture, as understood by Inuit and others, is the product of dichotomous perceptions of concepts such as traditional culture, and even Arctic. Inuit see themselves as the bearers of a unique adaptation, based on a continuing social relationship between people and animals (Nuttall 1992; Stairs and Wenzel 1992). In contrast, a Southerner often perceives the North to be a remote, pristine, and ecologically fragile landscape (e.g., Lopez 1986), resistant to only the most gentle of human intrusions. Inuit are implicitly expected to exist in this idealized environment and to measure up to the cultural standard of Robert Flaherty's 1921 film *Nanook of the North*.

To date sea level in Baffin Bay has not risen, nor have seal and caribou (*Rangifer tarandus*) populations followed retreating sea ice and tundra northward. However, Canadian Inuit are currently experiencing a "warming" unrelated to environmental factors, one in which Inuit are perceived by some Southerners to be the vanguard of environmental modernity. In this view, snowmobiles, rifles, and participation in the cash economy all symbolize the negative technoeconomic elements that affect our own environmental relations. Inuit are no longer considered original ecologists in natural equilibrium with the animals they harvest; as Steven Best (1985, personal communication) of the International Wildlife Coalition has stated, "They are just like us."

The effects of past climatic warming on Inuit must be explored before pursuing this new view of northern human ecological relations. However, one clear difference sets the present situation apart from that of Inuit cultural change during the last great Arctic warming. Modern Inuit are now being scrutinized by external groups that perceive little value in the central feature of Inuit culture: the capture and consumption of wildlife.

The Dorset-Thule transition

The last time that the North experienced climatic warming of a magnitude comparable to what is now forecast was ca. 1000 A.D., when the Canadian North entered the Neo-Atlantic (or Second Climatic) Optimum (McGhee 1970; Barry et al. 1977). The Arctic passed from a period of 400–500 years (the Scandic Period) in which circumpolar mean annual temperatures were 1°–2°C below present temperatures—a time of presumed environmental stress (McGhee 1970, 1972)—to one of climatic amelioration, with summer temperatures on the order of 2°C above present ones.

The warming resulted in less summer sea ice and longer open-water/ice-free summers in the eastern Canadian Arctic. This change in the duration and pattern of marine ice opened access to maritime habitat areas along Canada's Arctic coast to a variety of migratory small cetaceans and pinnipeds (narwhal [*Monodon monocerus*], beluga [*Delphinaptera leucas*], harp seal [*Phoca groenlandicus*], walrus [*Odobenus rosmarus*]), and, especially for Inuit cultural history, the bowhead whale (*Balaena mysticetus*).

The Neo-Atlantic Optimum also had far-reaching cultural effects on coastal prehistoric Eskimos. Foremost was the replacement of the 2,000-year-old Paleoeskimo Dorset culture by proto-Inupiat/Thule migrants from the Beaufort-Chukchi seas area (McGhee 1984; Maxwell 1985). The eastward spread of Thule culture from northern Alaska along the central Arctic coast of Canada was closely associated with the new ecological conditions. The new maritime conditions facilitated the movement of bowhead whales, the cornerstone of Thule Eskimo subsistence, into the formerly ice-closed waters of the eastern Arctic.

While Classic Thule whaling lasted only a few centuries (Andrews and Miller 1979), Thule technological innovations formed the material foundation of post-Thule Inuit, or Neo-Eskimo, society. Thule hunters encountered and presumably displaced the Tuniit (Dorset peoples) during their migration eastward. With some possible exceptions, their assimilation into Thule society was total,

---

Diamonds in the Far North

Canada's Northwest Territories (NWT) comprise 330 million hectares, but with a population of less than 60,000, it is one of the least densely populated regions of North America. With few roads connecting the small settlements, interaction among them is difficult, restricting the potential for economic development in the NWT. However, the region is rich in renewable and nonrenewable resources. Wildlife, forests, and marine life have been significant factors in the subsistence and culture of Native populations for thousands of years.

Nonrenewable resources have also played an important role in the economy and culture of the region. Native Yukon prospectors made some of the first finds that sparked the 1896–98 Gold Rush. In 1932, Eldorado became the NWT's first producing radium mine. Yellowknife boomed a year later with the discovery of gold. Ever since, the NWT have been mined for gold, copper, zinc, nickel, and most recently diamonds.

In the fall of 1991, diamonds were found near Lac de Gras by Dia Met Minerals. Since then approximately 20 million hectares have been staked by 22,000 claims that extend south from the Arctic coast to just north of Yellowknife. These claims are primarily on the mainland, but some exist on Baffin Island as well. The Native Inuit and Dene serve as guides, prospectors' assistants, cooks, and laborers. However, lack of formal education makes it difficult for them to attain the highly skilled jobs. Even with the low-skill labor positions, the shift from at least partial subsistence living to industrial cash-based employment is a major transition.

The NWT government demands extensive environmental review for mining operations, but there are many environmental concerns, including the potential for oil spills, contamination of whale and seal meat with PCBs, and growing deposits of garbage and industrial waste. At least one ton of ore must be removed, primarily by open-pit mining, to yield just one carat of diamonds, so large-scale mining would produce a dramatically altered landscape and surface hydrology. With little prospect for future restoration of ecosystems, the economic benefits of exploration and development must be weighed against the social and environmental costs.

Christy M. Parker

---

providing a dramatic example of what could result from an extreme warming in the Inuit cultural environment.

Although the archeological record as it pertains to eastern Arctic Dorset and Thule culture contacts is incomplete, several aspects seem clear. First, bowhead whale hunting appears to have had no precedent in Dorset culture. The ecological focus of Dorset resource subsistence activities approximate what others (McGhee 1972) have called a prehistoric variant of the Netsilik type of Inuit adaptation (Maxwell 1985).

Dorset culture is characterized as seasonally focused on hunting ringed seal (*Phoca hispida*) in winter, and exploiting migratory caribou herds supplemented by Arctic char (*Salvelinus alpinus*) fishing in summer. Walrus in Hudson Bay and musk-oxen (*Ovibos moschatus*) in the western Barrens and Arctic Archipelago were also important Dorset subsistence species. Dorset subsistence did not involve the exploitation of bowhead whales or other large cetaceans, and there is no evidence of whalebone tools or boat technology for whale hunting.

Artifactual data, including dog traction and skin boats (both *qajaq* and *umiaq* for whaling) indicate a technology shift during early Thule culture (e.g., Wenzel 1984; Maxwell 1985). A significant change is evident in Eskimo artifact types, especially toggling harpoon heads, compared with types known from Paleoeskimo Dorset culture. In addition, there was a major change in lithic (stone) raw material (from chert, quartz, and slate) and of artifact fabrication methods (flaking to grinding), first exhibited in Thule culture and then carried over through successive Neo-Eskimo variants until Inuit contact with Europeans. The Neo-Atlantic Optimum had as large an effect on the culture of northern Canada as it had on the biophysical environment.

Global warming and green trends

The Clyde area of Baffin Island (Figure 11.2) has short summers with cool temperatures (mean July temperature 4.5°C) and less than 60 open-water days (when

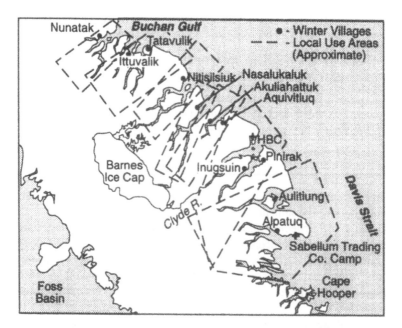

Figure 11.2. Baffin Island, with Clyde River and hunting camps indicated.

sea ice permits free use of small boats). During the 1980s, ice-out at Clyde village usually occurred between August 1 and 26. In the 1970s, there were three summers when breakup did not take place until the second week of September. In 1981 and 1982, however, open water began on July 16 and 18, respectively, the earliest such dates remembered by Clyde's oldest residents and recorded by weather observers.

But two unusually early breakups do not prove the existence of global warming. If such events meant no more than an early boating season, they would not necessarily have cultural importance. However, these early ice-outs were significant because Clyde hunters caught narwhal on the first day of openwater those two summers, four to six weeks earlier than average.

The practical significance of longer open-water whaling is that Clyde Inuit have a better chance of harvesting the 50 whales allocated to them under the government-mandated quota. There is also a cultural impact: more harvested narwhal provide more *maktaaq* (whale skin), an important food and source of vitamin C for the community. The 100 narwhal captured in 1990 and 1991 represent twice the annual rate of return (1950 kg to 3750 kg) normally obtained by the community (average of 24 narwhal per summer between 1956 and 1984) (Wenzel 1991). The value of the increased harvest was succinctly stated by a Clyde narwhal hunter (Ittuchuak, personal communication) in 1991: "If this is global warming, Inuit will love it."

The Neo-Atlantic warming first changed the ecology of the eastern Arctic and then affected Inuit culture. The driving force behind modern-day Inuit cultural change is an effort to alter southern perceptions of the Arctic, including the place of Inuit in northern ecosystems. This new view, espoused by members of the animal rights movement, does not perceive that new resources and technologies (snowmobiles, outboard engines) provide Inuit with an adaptive advantage (such as increased mobility) in the pursuit of traditional sociocultural goals (Wenzel 1991; Stairs and Wenzel 1992). Rather, eco-constructivists are concerned about the "fragile" Arctic environment, and have discovered that the North is as accessible and knowable as any place on earth (Lopez 1986). They also perceive that modern Inuit use of these tools results from cultural compromise and co-optation. Although increased concern about the long-term health of Arctic ecosystems is positive, eco-contructivist positions are based on ethnocentric definitions of environmental harm and good.

Seals, snowmobiles, and green concerns

Protests against seal harvests began in the mid-1950s. Awareness of seals as an environmental-political issue was far advanced by 1983, when the European Community officially banned seal imports, effectively closing the commercial market for sealskins. Hunting and market abolitionists felt that any attempt to isolate Inuit ringed seal harvesting from the protest's original focus on the industrial-scale killing of harp seal pups (Malouf 1986) off Canada's eastern coast would only derail the protest's momentum. The protest equated the ecological and cultural interests of Inuit with the commercial exploitation of harp seals (Goddard 1986; Wenzel 1991), despite efforts by scientists to inform the protest and its public about their differences.

Ringed sealskins, the cash mainstay of Inuit local economies between 1961 and 1983, rapidly lost value. A recasting of northern human-environment relations, and therefore of Inuit culture, had occurred. Inuit sealing was variously declared to be commercially motivated, removed from cultural tradition, and economically unnecessary. Even well after the sealskin ban, some activists have contended that because Inuit hunt with southern tools and technologies, buy imported foods, and are afforded modern housing by the Canadian government, the central role of wildlife to the Inuit no longer exists.

But considerable data suggest that wildlife must remain a part of contemporary Inuit life. Dietary (Borré 1989) and harvest (Usher and Wenzel 1989; Wenzel 1991) data from Inuit areas of Canada support the quantitative and nutritional importance of wild foods for Inuit (Table 11.1). The role of seal for the Inuit of Clyde River is particularly striking, with ringed seal (or *natsiq*) supplying the village with a large amount of high-quality protein (Table 11.2) at a monetary cost far below that of imported foods (Wenzel 1991).

Modern technology, especially the snowmobile, was adopted by Inuit to offset disruptions of traditional land use patterns experienced by virtually all Inuit

Table 11.1. *Composite harvest data for the four major Inuit regions of Canada.*

Region	Population	Total harvest, live weight (kg)	Species harvested (in order of importance by weight)
Inuvialuit (Mackenzie Delta, Banks Island, western Victoria Island)	6 permanent communities with 2,300 Inuit	393,000	Caribou, fish, beluga, waterfowl
Labrador	3 permanent communities with 1,670 Inuit	466,685	Ringed seal, caribou, harp seal, arctic fox
Nunavik Region (eastern Hudson Bay, Hudson Straight, Ungava Bay)	13 permanent communities with 6,500 Inuit	1,137,550	Ringed seal, fish, caribou, beluga
Nunavut (Baffin, Keewatin, and Kitikmeot regions of Northwest Territories)	26 permanent communities with 15,200 Inuit	5,430,500	Ringed seal, caribou, small cetaceans (narwhal, beluga), polar bear

Table 11.2. Mean seasonal ringed seal harvest by Clyde Inuit for the period 1981–1983.

Season	Number of seals harvested	Edible biomass of seals (kg)	Edible biomass from all sources (kg)	Biomass ratio of seal:all sources
Winter	932	21,436	43,016	0.50
Spring	984	22,640	37,400	.59
Summer	668	15,366	31,415	.49
Autumn	270	6,210	9,590	.64

from the 1950s onward as a result of government social policies (Wenzel 1986, Duffy 1988). Prior to the European Community boycott, ringed seal and seasonal game resources (polar bear [*Ursus maritimus*], caribou, fish, and small cetaceans) provided Inuit with high-quality foods and cash income for the purchase and operation of imported technologies. A striking impact of the sealskin boycott is that Inuit must now seek cash income outside traditional hunting activities. While access to traditional and imported resources was once available through traditional skills, Inuit must now possess abilities favored by non-Inuit. Working as an ecotourist guide or government clerk or driver provides cash, but reduces opportunities for subsistence hunting (Quigley and McBride 1987). It is unclear what this means to Inuit and their long-term adaptation to the environment. Certainly Inuit still hunt for much of their food, so there appears to be only slight erosion in the harvesting regime of communities like Clyde River. But this is only a superficial indicator. For example, it is evident that subtle changes are occurring in the cultural and physical pattern of Clyde harvesting, and social units associated with subsistence harvesting are now constructed differently.

While these phenomena may appear to be the effects of acculturation (e.g., Graburn 1969), they are associated with changes precipitated in the cash-subsistence relationship following the 1983 boycott. The weakening of kinship as a factor in hunt group affiliation and the centripetal movement of summer camps into areas adjacent to the main Clyde settlement are radical departures from the patterns of camp distribution and land use (Wenzel 1991) and task group formation (Wenzel 1981, 1994) that prevailed in the Clyde area into the early 1980s.

Before 1950, settlement and community patterns at Clyde were characterized by extended family (*ilagiit*) coherence and considerable mobility in summer (Figure 11.3). Summer land use involved the abandonment of winter village sites in May, whereupon each *ilagiit* used a number of temporary tent camps. In 1972, one summer family group (four households, 24 individuals) established 11 camps between late May and late August (88 days), occupying sites as briefly as two days and as long as 22 days. While some separation of households from the *ilagiit* might occur during the summer, especially in pursuit of caribou, the "fission-fusion model" (flexibility in residence, in which extended families split up and subsequently reassemble) said to typify Inuit settlement (Balikci 1968) was conspicuously absent from the Clyde area.

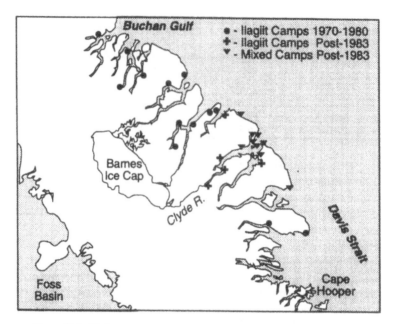

Figure 11.3. Clyde Inuit traditional ilagiit (extended family) areas, 1920–1945.

With the resettlement of the regional Inuit populations into centralized communities as directed by government policy in the 1950s and 1960s, the winter village and land use pattern at Clyde and other areas changed markedly. By 1969, only one of the Clyde region's eight *ilagiit* winter villages known to have been occupied in 1950 was active. Relocation meant localizing land use in winter around the main settlement area (mainly because of constraints imposed by dog traction), but traditional summer camp and land use activities remained relatively unaffected.

The ban on Canadian sealskin imports in 1983 immediately affected the economic state of eastern Arctic Inuit (Figure 11.4), for whom sealskins represented important income. The loss of stable, land-related access to cash through the sale of wildlife by-products (primarily sealskins, but also narwhal ivory and Arctic fox [*Alopex lagopus*] and polar bear skins) means that the costs of long-distance summer travel and large camp maintenance frequently exceed what many people can bear. Regional summer camp distribution has shrunk, leaving virtually all the traditional family areas in the region without summer occupation. The most actively and densely utilized area is now the Clyde Inlet-Inugsuin Fiord area (Figure 11.3) adjacent to the present settlement. A post-1983 adaptation has been three or four spring-summer "commuter camps," usually within 25 kilometers of the main village, from which wage earners travel daily from their jobs.

Prior to 1983, the social focus of multihousehold summer camps was the extended family, with economic responsibility and social authority mainly resting

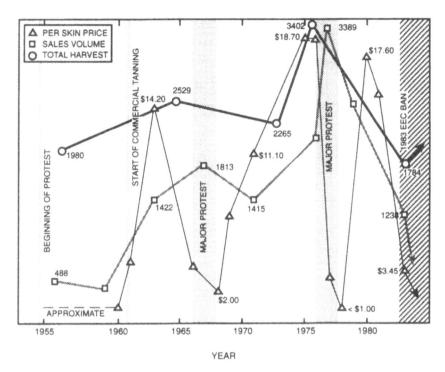

Figure 11.4. Clyde ringed sealskin prices, sales volume, and harvest, 1955–1983.

with adult hunters and their spouses. Since the sealskin boycott, the focus of these summer residential/task formations has shifted away from a core composed of close consanguinal male kinsmen (father and sons, brothers, uncles and nephews). Contemporary Clyde summer camps are increasingly based on households establishing residential association with one or more, often nonkin linked, male wage earners or with hunters with employed spouses (Figure 11.5). Such camp leaders have emerged because the current cash-economy condition (scarce wage and nonexistent wildlife income) makes it difficult for *ilagiit* to effectively organize the material resources needed for extended summer camp residence. The new leaders have emerged because they possess the harvesting equipment and can afford the high costs of summer camp, not because of knowledge about the environment and wildlife.

Inuit options for the future

As discussed above, the demise of the market for sealskins has eliminated an important source of Inuit hunter-generated monetary income. After 1984, when the full effects of the European Community sealskin ban began to be felt (at Clyde River in 1984–85, sealskin revenues amounted to $3,712 [Wenzel 1991], a

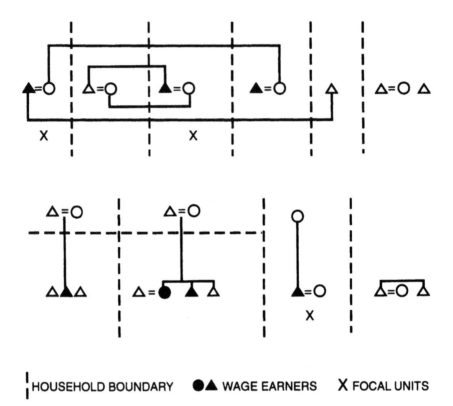

HOUSEHOLD BOUNDARY   ●▲ WAGE EARNERS   X FOCAL UNITS

Figure 11.5. Suluak camp composition, May 31–June 5, 1985 (upper figure), and Nuvuktiapik camp composition, June 19–July 26, 1985 (lower figure). Triangles indicate males, circles indicate females.

drop of $54,000 from 1980–81), the most direct effort to bridge the cash shortfall in the Inuit subsistence economy was made by Canada's Department of Fisheries and Oceans. This consisted of the offer of a quarterly "sealskin supplement," amounting to $6.50 per sealskin, to be added to the base commercial price in the village. But, the support program came into effect only after a hunter had sold 10 sealskins in a quarter, a qualification presumably set to avoid the appearance of a bounty.

In the mid-1980s, Clyde harvesters were receiving less than one-third of the 1980–81 per skin price for their first ten sealskins, with any further sales in a quarter fetching approximately 60% of the 1980–81 pelt value. However, the commercial price in Clyde had by then fallen to roughly one dollar. The Canadian government recognizes the high cost of Inuit subsistence capitalization and operations costs and the important contribution of harvesting to village food economies. The government support program has received participation mostly from committed older hunters, because younger hunters are unable to afford the entry price of harvesting. Thus the sealskin support program does not have the level of local enrollment anticipated by the government.

Other types of intervention have focused on ways to expand the local wage-labor and social-transfer sectors of the Inuit economy. The latter was done by liberalizing the qualifications for entitlement. Expansion of the wage sector has grown through the expansion of municipal services with funding provided through the local government program of the Northwest Territories. In Clyde River, employment by the hamlet, territorial, and federal governments grew from 15 of 34 positions in 1985 to 37 of 58 jobs in 1992, with the main additions being drivers, clerks, maids, janitors, and carpenter-mechanics. However, an increase in government jobs, coupled with more limited growth in other economic sectors, still left 60% of the Clyde Inuit adult population unemployed in 1991.

Ecotourism has shown potential for rapid growth and the ability to close the cash gap created by the European Community boycott (Nickels et al. 1991; Reimer and Dialla 1992). Inuit communities like Pangnirtung, Pond Inlet, and Bathurst Inlet are close to accessible national or territorial parks and nature reserves. Tourists are increasingly attracted to Nunavut in summer for mountain climbing, hiking, or birding and are an important source of seasonal employment (especially for hunters willing to forgo harvesting to be tourist guides).

One cost of seeking tourist dollars is the curtailment of hunting in areas frequented by visitors, out of fear that any sight of consumptive wildlife use will adversely affect this new trade (Nickels 1992; Greken 1994). Tourist data from Pond Inlet confirm that such fears are well founded. As one visitor said, "Although I understand the Native peoples of the North must hunt to live, I don't care to witness it" (anonymous, 1990, personal communication). With federal and territorial authorities responding affirmatively to southern calls for expanded conservation in the North, it is apparent that some Inuit communities may find themselves in the ironic position of seeking tourism dollars at the cost of declining subsistence harvesting. In communities like Clyde River, community discussions of tourism focus on the relative monetary benefits accrued by a few individuals versus the costs in time, energy, and possibly food involved in shielding visitors from the visual realities of harvesting.

Summary

The Neo-Atlantic warming changed but also reinvigorated Inuit society. Whether the present warming will be as kind is difficult to say. It is clear that the current biosocial situation in the North may have a greater impact on Inuit culture than the early transition from Paleoeskimo to Thule culture. An era that began with the expansion of Inuit culture through its adaptation to large whale hunting may end with the demise of Inuit harvesting because of the environmental imperatives of the South.

An evaluation of the cultural biases of North-focused environmentalism (Nelson 1993) could help ameliorate ongoing biosocial impacts on Inuit. A positive indication is the recent cooperation between the Inuit Circumpolar Conference and international environmental organizations over the Arctic component of the World Conservation Strategy. However, misrepresentations of Inuit wildlife use and subsistence, the anti-sealing campaign of the 1970s and 1980s, and the present objections of some environmentalists to changes in the Marine Mammal Protection Act that might benefit Inuit have left a lasting impression on the Inuit of Clyde and other small, wildlife-dependent communities. They clearly wonder if any rapprochement can ever take place.

References

Andrews, J. T., and G. Miller. 1979. Climatic change over the last 1000 years, Baffin Island, N.W.T. In *Thule Eskimo culture: An anthropological retrospective*, 541–554. Archaeological Survey of Canada Mercury Paper 88. Ottawa: National Museum of Man.

Balikci, A. 1968. The Netsilik Eskimos: Adaptive processes. In *Man the hunter*, edited by R. B. Lee and I. DeVore, 78–82. Chicago: Aldine.

Barry, R. G., W. Arundale, J. T. Andrews, R. Bradley, and H. Nichols. 1977. Environmental change and cultural change in the eastern Arctic during the last five thousand years. *Arctic and Alpine Research* 9:193–210.

Borré, K. 1989. A bio-cultural model of dietary decision making among North Baffin Island Inuit: Explaining the food consumption of Native Canadians. Ph.D. Dissertation, Department of Anthropology, University of North Carolina, Chapel Hill, NC.

Duffy, R. Q. 1988. *The road to Nunavut: The progress of eastern Arctic Inuit since the Second World War*. Montreal: McGill-Queen's Press.

Flaherty, R. 1921. *Nanook of the north*. Revillion Fréres.

Goddard, J. 1986. Out for blood. *Harrowsmith* X:29–37.

Graburn, N. H. H. 1969. *Eskimos without igloos: Social and economic development in Sugluk*. Boston: Little, Brown and Co.

Greken, J. 1994. Understanding the community-level impacts of tourism: the case of Pond Inlet. M.A. thesis. Department of Geography, McGill University, Montreal, Canada.

Gross, P. R., and N. Levitt. 1994. *Higher superstition: The academic left and its quarrels with science*. Baltimore: Johns Hopkins University Press.

Lopez, B. 1986. *Arctic dreams: Imagination and desire in a northern landscape*. New York: Scribner.

Malouf, A. 1986. *Seals and sealing in Canada*. Report of the Royal Commission on Seals and the Sealing Industry in Canada. Ottawa: Department of Fisheries and Oceans.

Maxwell, M. 1985. *Prehistory of the eastern Arctic.* New York: Academic Press.
McGhee, R. 1970. Speculations on climatic change and Thule culture development. *Folk* 11/12:173–184.
McGhee, R. 1972. Climatic change and the development of Canadian Arctic cultural traditions. In *Climatic change in Arctic areas during the last ten thousand years,* edited by Y. Vasari, 39–57. Acta Universitatis Ouluensis Series A, Scientiae Rerum Naturalium 3, Geologica 1. Oulu, Finland: University of Oulu.
McGhee, R. 1984. Thule prehistory of Canada. In *Handbook of North American Indians, Vol. 5—Arctic,* edited by D. Damas, 369–376. Washington, DC: Smithsonian Institution.
Nelson, R. 1993. Understanding Eskimo science. *Audubon* 95:102–104.
Nickels, S. 1992. Northern conservation and tourism: The Perception of Clyde River Inuit. M.A. Thesis. Department of Geography, McGill University, Montréal, Canada.
Nickels, S., S. Milne, and G. W. Wenzel. 1991. Inuit perceptions of tourism development: The case of Clyde River, Baffin Island, N.W.T. *Etudes/Inuit/Studies* 15:157–69.
Nuttall, M. 1992. *Arctic homeland: Kinship, community and development in Northwest Greenland.* Toronto: University of Toronto Press.
Quigley, N., and N. McBride. 1987. The structure of an Arctic microeconomy: The traditional sector in community economic development. *Arctic* 40:204–211.
Reimer, G., and A. Dialla. 1992. *Community based tourism development in Pangnirtung, Northwest Territories: Looking back and looking ahead.* Yellowknife: Department of Economic Development and Tourism, Government of the Northwest Territories.
Stairs, A., and G. W. Wenzel. 1992. "I am I and the environment:" Inuit hunting, community, and identity. *Journal of Indigenous Studies* 3:1–12.
Usher, P. J., and G. W. Wenzel. 1989. Socio-economic aspects of harvesting. In *Keeping on the land: A study of the feasibility of a comprehensive wildlife support programme in the Northwest Territories,* edited by R. Ames, 1–62. Ottawa: Canadian Arctic Resources Committee.
Vibe, C. 1967. Arctic animals in relation to climatic fluctuations. In *Meddelelser om Gronland,* Vol. 170(5). Copenhagen: C.A. Reitzels Forlag.
Wenzel, G. W. 1981. *Clyde Inuit adaptation and ecology: The organization of subsistence.* Canadian Ethnology Service Mercury Paper 77. Ottawa: National Museum of Man.
Wenzel, G. W. 1984. Archaeological evidence for prehistoric Inuit use of the sea ice environment. In *Sikumiut: The people who use the sea ice,* edited by A. Cooke and E. Van Alstine, 41–60. Ottawa: Canadian Arctic Resources Committee.
Wenzel, G. W. 1986. Canadian Inuit in a mixed economy: Thoughts on seals, snowmobiles, and animal rights. *Native Studies Review* 2:69–82.
Wenzel, G. W. 1991. *Animal rights, human rights: Ecology, economy and ideology in the Canadian Arctic.* London: Belhaven Press.

Wenzel, G. W. 1994. Recent change in Inuit summer residence patterning at Clyde River, East Baffin Island. In *Key issues in hunter-gatherer research*, edited by E. S. Burch and L. Ellanna, 289–308. Oxford: Berg Publishers. In press.

Woodbury, A. C. 1985. Eskimo and Aleut languages. In *Handbook of North American Indians, Vol. 5—Arctic*, edited by D. Damas, 49–63. Washington, DC: Smithsonian Institution.

Part IV

Natural Resources and Human Institutions in a Dynamic Environment

It is ironic that northern rural people, who have had almost nothing to do with imminent climate change, may be among the world's most vulnerable groups. They have little impact on large-scale changes in their environment, but must adjust to whatever changes are imposed from "outside."

We now begin our search for solutions. Just as the causes of potential climate change are deeply embedded in human social structure and culture, so too are the processes by which people will cope with these changes. In the very long term, climate change could result in a more biologically productive North, bringing with it many opportunities. It is likely, however, that the next century will be characterized by more problems than opportunities, as natural and social systems undergo unprecedented disruption.

Assertive social action—prior to major impacts—to prepare for these potential changes is preferable to laissez faire approaches with short-term responses to crises. Unfortunately, the uncertainties associated with climate change make systematic planning difficult. A first step may be to discuss relevant issues in open, interdisciplinary forums. This should be done at all levels: local, regional, national, and international. We also need to evaluate the effectiveness of existing social institutions in providing mechanisms for adaptation to change.

This section begins that process. The collective message is one of skepticism regarding the ability of existing institutions to cope with future challenges. There are some formidable barriers: interagency conflict, lack of interdisciplinary scientific coordination, cross-cultural communication problems, and the slow pace at which conflict resolution is being instituted. Fortunately there are some productive strategies for building institutions that will be able to address dynamic (and unpredictable) biosocial issues of the future. The following chapters are not an exhaustive discussion of social processes and institutions. Many other topics could be examined. However, they are a good start toward understanding and managing complex environmental and social issues in the North.

12

Global warming and conflict management: Resident native peoples and protected areas

Patrick C. West

There is a growing comparative case study literature on relations between resident peoples and protected areas, a considerable amount of research literature on environmental conflict management, and a voluminous if inconsistent literature on global warming. However, the intersection of these topics in relation to conflict management decision-making processes has not been addressed. A provisional orientation is needed to identify issues for discussion and debate.

> The greatest and most troubling conflicts are not between good and evil but between good and good. In conflict between good and evil the solution is simple—seek the triumph of good over evil. But in the conflict between good and good the balancing of conflicting moral imperatives is painful and trying, and without clear implications for a correct course of action. The resident peoples issue (vis-à-vis protected area conflicts) is clearly in this latter category. (West and Brechin 1991, p. xix)

A good vs. good conflict is emerging with regard to social policy relevant to potential rapid climate change at northern latitudes. An additional situation is emerging in which the achievement of each side's good objectives depends on the other. Because we are dealing with a situation in which good conflicts with good, it is important to create effective decision-making processes that incorporate institutions for conflict management.

Two sociological premises

When science debates, society is alternately frightened, confused, or bored. Thus an understanding of scientific vs. societal uncertainty regarding climate change has two important implications. First, there are two processes occurring: (1) a scientific debate regarding the validity of climate change predictions, and (2) a social reaction to the scientific debate, including partisan debate with advocates using scientific experts to support competing positions. Until undeniable evidence is introduced that climate is rapidly changing, it is highly unlikely that society will act. This delayed reaction will result in a contentious and reactive decision-making process that will occur over a comparatively short time.

Assuming rapid climate change occurs, social institutions are needed that are responsive to a potential impending crisis. If general circulation models are correct, then the northern latitudes will have much greater temperature increases than southern latitudes (Chapter 2). Because society is not likely to respond much in advance due to scientific uncertainty, social institutions for conflict resolution and cooperation should be constructed now in anticipation of environmental and social changes. Some mechanisms for resolution of problems will then be in place if they are needed because of climate change or other stresses.

There are many reasons other than climate change for building institutions for conflict resolution. There is a growing consensus in the conservation community that excluding the participation of resident peoples is counterproductive to conservation, aside from the moral and political implications of marginalizing these social groups (Marks 1984; Brechin and West 1990; West and Brechin 1991; Gray 1993).

For example, social instability in Togo, combined with repressive government policies toward displaced local peoples, led to a rampage that destroyed much of the flora and fauna of Togo's national parks (Lowry and Donahue 1994). In the face of such object lessons, there is a new and growing interest in merging conservation and development in a way that requires strengthening institutions of cooperative management and conflict management for protected area decision making (Bidol and Crowfoot 1991; East 1991; Weaver 1991).

Fortunately, northern North America is located within secure constitutional democracies. It is also fortunate that innovative institutions for conflict management and cooperative management are evolving in and around the new national parks of northern North America. Contrast this circumstance with the developing world, where local peoples are often excluded from decision-making processes, perhaps displaced, and sometimes virtually destroyed [for example, in Uganda (Calhoun 1991), India (Raval 1994), and Benin (Agbo et al. 1993)]. Compare the paralyzing effect of the intimidation felt by resident peoples toward local officials in Benin, West Africa—threatened with starvation—following their displacement from national parks, to the successfully aggressive stance of the Havasupai of Grand Canyon National Park toward the U.S. National Park Service (Hirst 1985; Hough 1989; Simsik et al. 1993).

Many challenges remain in North America. For example, while the park managers of Wood Buffalo National Park (Alberta and Northwest Territories) consider the current cooperative management and conflict management regimes fair and effective (East 1991), some aboriginal leaders strongly disagree. And while new legislation is supposed to ensure a more level playing field for resident peoples, recent evaluations in Gates of the Arctic National Park (Alaska) show these institutions to be deficient in providing effective mediation of disputes between Native peoples and the National Park Service (Mathews 1993).

Similar situations exist in other regions that also have advanced democratic states. Conservationists tout Kakadu National Park in Australia as a model of cooperative management and participatory conflict resolution (Hill 1983). However, decisions at the policy-making level are monopolized by national institutions, with little or no participatory conflict management from local aboriginals (Weaver 1991). Organizations such as the World Wildlife Fund may seem enlightened in terms of integrating conservation and development, particularly in areas such as Annapurna, Nepal (Bunting et al. 1991). However, actual implementation can result in inequities in participation and distributive outcomes for indigenous peoples. Furthermore, World Wildlife Fund projects in Madagascar emphasize conservation more than participatory development and emphasize basic needs compensation more than participatory rural development (Hough 1994). This is largely due to the inherent "organizational culture" of conservation agencies. There is clearly reason for concern about the viability of current conflict management systems, even within the most favorable organizational contexts and without the complications of climate change.

Conflicts between Native peoples and protected areas under conditions of climate change

Climate change and its impacts will not ensue from the actions of either Native peoples or protected area managers, yet they will both be left to deal with the consequences. This situation is different from typical labor or environmental mediations. If labor does not get a raise, management can be held responsible; if old-growth forest is cut, it is seen as the fault of either the timber companies or the forest management agency.

Resource managers and Native communities may see climate change as a common enemy, and a coalition on one front may mute or reduce conflicts on the other fronts. For example, banding together to tackle the challenges of climate change may reduce conflicts over the use of protected area resources by Native peoples. However, other outcomes are equally possible.

Social conflict stemming from the impacts of climate change is most often perceived as a conflict between winners and losers. For example, higher temperatures might produce longer and warmer summer seasons in the Far North (Chapter 2), leading to greater biomass production (Chapter 5). If Native peoples have been demanding access to x biomass (mass of living plants and animals), there might now be $x + 10$ biomass to which they want access (potential global warming winners). However, they might be resisted by protected area management—and related environmental constituencies and interest groups—who do not want humans to disrupt the ecosystem. In some cases, climate change might be viewed as an unnatural perturbation; therefore, resource managers might see themselves as losers, and demands by potential winners for more resources would add to the tension.

Alternatively, permafrost changes in the Arctic and subarctic interior and expansion of black spruce (*Picea mariana*) tree cover farther north into the tundra (Chapter 5) might change the migration routes of caribou (*Rangifer tarandus*) (Chapter 6). Animals that formerly migrated through protected areas might be dispersed beyond those areas, resulting in greater contention between park management, environmentalists (losers), and Native peoples (winners), who would now have greater access to caribou. Protected area managers (losers) might, in turn, exercise their power in an effort to control the shifting balance.

Harvey (1991) points to the potential need to extend protected area boundaries to protect wildlife if changing ecosystem parameters move key species beyond the boundaries. Predictions have been made that a warmer climate may drive caribou out of Prince Albert National Park (Saskatchewan), necessitating the extension of park limits (Wall 1989; Harvey 1991). If this were to occur, it might negatively affect Native peoples' subsistence efforts while providing new protection for wildlife. A climate change winner could become a loser, and a loser could become a winner, resulting in a new and increased state of conflict between the emerging winner (the park system) and loser (Native peoples). Given the significant power of the state over Native peoples and the animosity that would inevitably accompany extension of bureaucratic authority, this conflict would be particularly difficult to manage.

What about conflicts between winners and winners, or losers and losers? Of the two, a loser/loser conflict caused by resource scarcity seems the most likely. Climate change could alter the distribution and abundance of some species of flora or fauna (Chapters 4 and 5). Indigenous peoples may want to continue harvesting these resources under former agreements of cooperative management, but the protected area management may apply new restrictions. Continued harvest of a particular species by Native peoples may actually drive that species more rapidly toward extinction within protected area boundaries, resulting in conflict between two groups of global warming losers.

Win/win outcomes are usually considered the best result in conflict management; indeed, they are the goal of much conflict management theory and practice. However, dissension between climate change winners may also engender considerable conflict. Suppose a warmer climate drives greater numbers of nature-oriented tourists from the increasing summer heat of the South to parks in Alaska. Cooperative management agreements grant economic benefits of ecotourism to Native peoples, but the National Park Service may want to recoup tourism losses in the South by remonopolizing economic benefits of ecotourism. A similar situation already exists in park management in China (Machlis and Johnson 1987) and Benin (West et al. 1993).

Conflict over surpluses can be as intense as struggles over scarce resources (Weber 1968; West 1975). Substantive conflicts could emerge over winner/winner and loser/loser situations as frequently as over the more common winner/loser circumstances. Which of these tendencies will be most common when climate change occurs in ecosystems of the Far North? Unfortunately, social conflict can be only generally anticipated until biological consequences are evident.

Water, Electricity, and the Cree

Hydro-Quebec's development of James Bay has inspired controversy since its conception in the early 1970s. Hydro-Quebec views the vast rivers of this region as potential power waiting to be harnessed by dams and hydroelectric facilities. This hydrologic resource will be used to promote Quebec's energy sovereignty, while the sale of excess power will reduce the province's deficit. However, to the Cree, who have subsisted on the land for 5,000 years, these rivers and forests are their life and culture.

The completion of the La Grande Dam in 1982 (Phase I), which inundated nearly 1.2 million hectares of forest and altered the flow of three rivers, caused the Cree to seriously consider the effects of future resource development on their land. The question remains whether additional power supplied by completion of Phase II and Phase III in the James Bay region is truly needed. The Cree now promote conservation rather than increased development, as they face social and environmental consequences that penetrate deep into the core of their society.

The construction of the La Grande Dam brought a rapid change in Cree lifestyle. Jobs took Natives away from traditional hunting practices, while increased income provided amenities such as televisions, snowmobiles, and electric lights. However, there have also been increases in drug abuse, alcoholism, and crime. In just two decades the Cree have been forced to adjust their lifestyles in response to dramatic changes in socioeconomic conditions. Many Cree now ponder whether future generations will remember their heritage.

The James Bay Project has diverse environmental consequences for the Far North. The project targets 19 river drainages for modification by more than 200 dams and dikes. Huge areas of spruce and hardwood forests will be lost under water, decreasing habitat for birds, mammals, fish, and amphibians. Scientists have determined that decaying vegetation under water is releasing methyl mercury, a neurotoxin. The methyl mercury is consumed by fish, which are later eaten by predatory birds, bears, and most notably the Cree. This unforeseen toxicological insult, accompanied by social and environmental impacts, has turned the Cree even more strongly against Hydro-Quebec and the provincial government. It is unlikely that the Cree will allow future resource development in their homelands without a more comprehensive assessment of potential consequences.

Christy M. Parker

Preexisting conditions

Conflict caused by climate change will be superimposed on preexisting conditions that vary widely from region to region. In Uganda, for example, the displacement of the Ik people resulted in their first sudden exposure to the outside world, coupled with a swift conversion from being hunter-gatherers to being dryland farmers. These conditions led to intense conflict with government officials and a disastrous impact on the culture of the Ik (Calhoun 1991). In marked contrast, a government request for local people to move out of a national park in Swaziland was met with cooperation because the local people were already farmers, were familiar with the outside culture, and were offered better land (Ntshalintshali and McGurk 1991). The concurrence of King Suhbbuza gave this process further legitimacy and minimized conflict.

What are the characteristics of biophysical resources and cultures in North America that represent important differences in pre-conditions of conflict and conflict management institutions? Comparative research is needed to answer this question. Conditions may be quite different in the diverse biological and social systems of the Arctic National Wildlife Refuge (northern Alaska), Wrangell-St. Elias National Park (southern Alaska), and Wood Buffalo National Park (Alberta and NWT).

Conflict management strategies

Potentially heightened states of conflict combined with the likelihood of reactive decision making suggest that there are some specific innovations in conflict management institutions that should be considered. It is beyond the scope of this chapter to review the full array of steps and processes involved in standard conflict management practices (Manring et al. 1990; Bidol and Crowfoot 1991). It is important here to suggest modifications in these standard processes that would be needed for optimal effectiveness under conditions of climate change. Given the combination of potentially heightened conflict and minimal lead time, it may be most effective to redefine the process of Alternative Environmental Conflict Management (AECM).

AECM has been touted as an alternative to legal court institutions for mediating conflicts. For these processes to work outside of the court system, however, there must be incentives for all parties to participate and a level of conflict that has not gone too far. Howard Belman (personal communication, University of Wisconsin, Madison, 1980) suggests that AECM will be successful only if a conflict has not already become too complicated to be resolved. In fact, the overlay of multiple levels of conflict caused by impacts of climate change, combined with reactive decision making, may push the intensity of conflicts to a level at which AECM cannot function.

There is a growing precedent for court-based conflict management in conflicts involving natural resources and Native peoples. During an Indian fishing rights case concerning Michigan's Great Lakes, out-of-court negotiations broke down, and the settlement process went to court. While some decisions were made by the judge (e.g., Native fishing rights), the issue of allocating fishing quotas was bound over to a special form of AECM known as the Special Master Process. It is similar to AECM except that, rather than being an alternative to court process, it operates within court-based institutions and provides a process similar to binding arbitration.

The "special master" is selected by the judge with the approval of all key stakeholders. Within this structure, regular mediator role strategies are employed by the special master to bring the parties together. If they can agree, fine; if they cannot, the judge will impose a settlement (i.e., the mediation is bounded, as in binding arbitration). This creates an incentive for each party to compromise because they are aware that the judge could impose a decision far less satisfactory than what they could achieve by cooperating in mediation with the special master.

A recent conceptual effort by sociologists links the importance of predictive social impact analysis (PSIA) to effective conflict management processes (Manring et al. 1990). The Sylvania Wilderness (Michigan) is an area where intense conflicts between resident peoples, wilderness managers, and related interest groups could have been avoided, if procedures for predicting tourist demand had been employed in a PSIA process linked with conflict management processes. Predicting social impacts from alternatives for protected area management is difficult due to unforeseen events and contingencies (Geisler 1993). Therefore, an ongoing monitoring approach to PSIA in relation to conflict management decision making is needed for protected areas. This argument holds even greater weight when the considerable uncertainties of the impacts of climate change are considered. Ongoing conflict management for adaptive management becomes increasingly important as unanticipated consequences of climatic warming emerge.

In addition to adaptations of conflict management models due to the extra stress of global warming, there are a number of other adaptations that are needed because of the cross-cultural nature of these mediations. For example, a scientific presentation to Native peoples about the potential threats of a warmer climate on ecosystems may not be clearly understood. Similarly, there may be indigenous knowledge (e.g., about changes in the Inuit 25 kinds of snow; see Table 22.1) that may not be recognized by modern science, but which is relevant to local understanding of climatic trends and valuable in monitoring environmental conditions (Chapter 15). There is a need to communicate in a cross-culturally sensitive way that translates not just between languages but between cultures (Bidol and Crowfoot 1991). There may not be comparable words in Native languages for terms such as carbon dioxide. By the same token, some concepts in indigenous languages may be difficult to translate into English or French. Therefore, multicultural communications will be improved only if cross-cultural principles are understood by all parties (Bidol and Crowfoot 1991).

Conflict management mechanisms may be varied among different cultures. For example, based on principles in the anthropology of law, rational legal law-based adjudication for resolving conflicts may not be appropriate in traditional social systems (Bidol and Crowfoot 1991). A careful understanding of differences between modern bureaucratic systems of conflict management and informal, traditional systems is critical to developing a consensus-based process of managing conflicts. Lack of attention to this critical issue could doom the mediation process to failure.

Summary

Can a more equitable balance of conservation objectives and development for Native peoples in protected areas be preserved during a transition through the impacts of climate change by employing conflict management institutions in decision making? Yes, it can. Institution building for conflict management must go beyond the successful examples of cooperative management regimes in parks such as Wood Buffalo (East 1991). While the state-of-the-art of conflict management will no doubt change as the impacts of climate change become more evident, a blueprint is presented here of how these techniques might be institutionalized.

Merging conservation and development is difficult under any circumstances. It will be more difficult under conditions of climate change. Despite the uncertainties that cloud our view of the future, institutions that integrate cooperative management regimes (Chapters 13 and 14) with conflict resolution components makes good sense for the managerial challenges of today. They could be indispensable in the future.

References

Agbo, V., N. Sokpon, J. Hough, and P. West. 1993. Population-environment dynamics in a constrained ecosystem in Northern Benin. In *Population-environment dynamics: Ideas and observations*, edited by G. Ness, W. Drake, and S. Brechin, 283–300. Ann Arbor: University of Michigan Press.

Bidol, P., and J. Crowfoot. 1991. Toward an interactive process for siting national parks in developing nations. In *Resident peoples and national parks: social dilemmas and strategies in international conservation*, edited by P. West and S. Brechin, 283–300. Tucson: University of Arizona Press.

Brechin, S., and P. West. 1990. Protected areas, resident peoples, and sustainable conservation: The need to link top-down with bottom-up. *Society and Natural Resources* 3:77–79.

Bunting, B. W., M. N. Sherpa, and M. Wright. 1991. Annapurna Conservation Area: Nepal's new approach to protected area management. In *Resident peoples and national parks: Social dilemmas and strategies in international conservation*, edited by P. West and S. Brechin, 160–172. Tucson: University of Arizona Press.

Calhoun, J. B. 1991. The plight of the Ik. In *Resident peoples and national parks: Social dilemmas and strategies in international conservation*, edited by P. West and S. Brechin, 55–60. Tucson: University of Arizona Press.

East, K. M. 1991. Joint management of Canada's northern national parks. In *Resident peoples and national parks: Social dilemmas and strategies in international conservation*, edited by P. West and S. Brechin, 333–345. Tucson: University of Arizona Press.

Geisler, C. 1993. Rethinking SIA: Why ex ante research isn't enough. *Society and Natural Resources* 6:327–338.

Gray, G. G. 1993. *Wildlife and people: The human dimensions of wildlife ecology.* Chicago: University of Illinois Press.

Harvey, L. 1991. Climate change: Warming to the challenge. In *The state of Canada's environment*, edited by Canada Ministry of Environment, 22–1 + 22–28. Ottawa: Environment Canada.

Hill, M. 1983. Kakadu National Park and the aboriginals: Partners in protection. *Ambio* 12:158–167.

Hirst, S. 1985. *Havasuw 'Baaja: People of the blue green water.* Supai, AZ: The Havasupai Tribe.

Hough, J. 1989. National parks and local people relationships: Case studies from Northern Benin, West Africa, and the Grand Canyon, USA. Ph.D. dissertation. University of Michigan, Ann Arbor, MI.

Hough, J. 1994. Institutional constraints to the integration of conservation and development: A case study from Madagascar. *Society and Natural Resources* 7:119–124

Lowry, A., and T. Donahue. 1994. Parks, politics, and pluralism: The demise of Togo's national parks. *Society and Natural Resources* 7:321–329.

Machlis, G., and K. Johnson. 1987. Panda outposts. *National Parks* 61:14–16.

Manring, N., P. West, and P. Bidol. 1990. Social impact assessment and environmental conflict management: Potential for integration and application. *Environmental Impact Assessment Review* 10:253–264.

Marks, S. 1984. *The imperial lion: Human dimension of wildlife management in central Africa.* Boulder, CO: Westview Press.

Mathews, V. T. 1993. A case study of a homeland and a wilderness: Gates of the Arctic National Park and the Nunamiut Eskimos. M.S. thesis. University of Alaska, Fairbanks, AK.

Ntshalintshali, C., and C. McGurk. 1991. Resident peoples and Swaziland's Malolotja National Park: A success story. In *Resident peoples and national parks: Social dilemmas and strategies in international conservation*, edited by P. West and S. Brechin, 61–67. Tucson: University of Arizona Press.

Raval, S. R. 1994. Wheels of life: Perceptions and concerns of the resident peoples for Gir National Park in India. *Society and Natural Resources* 7:305–320.

Simsik, M., V. Agbo, and N. Sokpon. 1993. Farmer participation in pre-project assessment of receptivity to agroforestry: A feasibility study in the village of Tanougou, Republic of Benin, West Africa. Natural Resources Sociology Research Lab, Technical Report 7. Ann Arbor: School of Natural Resources and Environment, University of Michigan.

Wall, G. 1989. *Implications of climate change for Prince Albert National Park, Saskatchewan. Climate Change Digest* 89–03. Downsview, Ontario: Atmospheric Environment Service, Environment Canada.

Weaver, S. 1991. The role of aboriginals in the management of Australia's Cobourg (Gurig) and Kakadu National Parks. In *Resident peoples and national parks: Social dilemmas and strategies in international conservation*, edited by P. West and S. Brechin, 311–332. Tucson: University of Arizona Press.

Weber, M. 1968. *Economy and society: An outline of interpretive sociology*, edited by G. Roth and C. Wittich. Translated by G. Roth. New York: Bedminster Press.

West, P. 1975. Social structure and environment: A Weberian approach to human ecological analysis. Ph.D. dissertation, Yale University, New Haven, CT.

West, P., and S. Brechin, eds. 1991. *Resident peoples and national parks: Social dilemmas and strategies in international conservation*. Tucson: University of Arizona Press.

West, P., V. Agbo, N. Sokpon, and M. Simsik. 1993. Villager perceptions and attitudes towards eco-tourism in Tanougou, N. Benin (West Africa). Natural Resources Sociology Research Lab Technical Report 8. Ann Arbor: School of Natural Resources and Environment, University of Michigan.

13

Comanagement of natural resources: Some aspects of the Canadian experience

Peter J. Usher

How can people who are few in number, and living in scattered and isolated villages far from the centers of power, effectively limit, mitigate, and adapt to the effects of climate change on their lands and resources? One approach is to form alliances and act cooperatively with others who may have similar needs and interests. Does the emerging system of resource comanagement in Canada offer a model for this?

Origins and development of comanagement

Comanagement of natural resources refers to institutional arrangements whereby governments and aboriginal entities (and sometimes other parties) enter into formal agreements specifying their respective rights, powers, and obligations with respect to the management and allocation of resources in a particular area. Comanagement is a form of power sharing, although the relative balance and the means of implementation vary from case to case.

Twenty years ago, all Canadian governments regarded their authority with respect to land and resource management as unlimited, except (in the most minimal way) by signed Indian treaties. The origins of comanagement are in crisis and struggle as a result of aboriginal land claims, real or perceived resource depletion, and adverse impacts of development and the need to mitigate them. Existing comanagement arrangements are a compromise. They do not provide for complete aboriginal self-determination or self-management, but they do place significant limits on government management authority.

For aboriginal people, comanagement is one solution to the progressive encroachment and restriction on the use of customary lands and resources; to harvest disruption; and to the loss of social, cultural, and economic values. For governments, comanagement is a means of enlisting hunter cooperation to ensure conservation, an alternative to deploying draconian and expensive enforcement measures with limited success. Some observers have suggested that comanagement arrangements offer a potential bridge between indigenous and state systems of knowledge and management (Usher 1987; Osherenko 1988). The record of achievement in this regard is mixed, but unquestionably they have provided a forum or venue for continuing negotiation over matters crucial to both aboriginal peoples and governments.

The need for government authority and initiative to be more client-responsive is a general social and political trend. This trend has been strongly reinforced in Canada, with respect to aboriginal peoples, by recent constitutional and legal developments such as the *Constitution Act, 1982,* and the Supreme Court's judgment in *R. v. Sparrow.* That decision provided a further impetus for comanagement, because consultation has become one of the key tests of constitutionally acceptable limitations on aboriginal harvesting rights. The boards also provide a useful "single window" for governments to deal with specific resource issues.

Some comanagement arrangements constitute an ad hoc, and possibly temporary, policy response to crisis. These include the oldest and most widely known, such as the Beverly-Qamanirjuaq Caribou Management Board (BQCMB), as well as many recent and geographically restricted arrangements, such as the Wendaban Stewardship Authority in northeastern Ontario, recently terminated unilaterally by the Government of Ontario. Others are central elements of comprehensive land claims agreements. These agreements (in effect, modern treaties), negotiated between aboriginal nations or peoples and the Government of Canada on the basis of aboriginal title or treaty rights to traditional lands, are protected under section 35 of the Constitution. They thus provide a much stronger guarantee and enhancement of the aboriginal interest than the ad hoc comanagement arrangements.

There are three important features of the claims-based regimes that are critical to the successful implementation of comanagement. First, the comanagement structures and their mandate, objectives, and mode of operation, are themselves negotiated. This is very different from inviting people to sit on a body whose parameters have already been determined unilaterally. Second, aboriginal members of claims-based boards are politically accountable representatives of one of the parties to an agreement, not simply stakeholders or users, as is the case on the ad hoc boards. In some of the latter type of boards, (including the well-known BQCMB), only governments are signatories to the management agreement. The rights and powers of users are specified but not guaranteed; they are granted by governments and do not constitute a recognition of existing rights. Third, only the claims-based arrangements are permanent. The ad hoc arrangements are in place for only a limited period, subject to discretionary renewal and funding.

What claims-based comanagement provides

Several comprehensive land claims agreements providing for comanagement have been negotiated since 1975, mostly in northern Canada. These include the

James Bay and Northern Quebec Agreement (JBNQA), the Inuvialuit Final Agreement (IFA), the Nunavut Final Agreement (NFA), the Yukon Umbrella Final Agreement (YUFA), and the Gwich'in and Sahtu Agreements in the lower Mackenzie Valley. In each case, a new structure is created: a board whose members are appointed in equal numbers by government and beneficiaries. The responsibilities and powers of the boards fall into two main spheres: allocation, in which they have actual decision-making power, and management, in which technically they have only advisory roles. Allocation and licensing are generally delegated to the boards and the local harvester organizations, and management for conservation remains the prerogative of governments, although there is substantial variation with respect to the latter. In the JBNQA, the Cree and Inuit have consultative roles of limited effectiveness, whereas in the IFA, the comanagement bodies effectively determine conservation. In the case of the NFA, for example, the Nunavut Wildlife Management Board (NWMB) may approve management plans; the establishment of conservation areas and management zones; and the designation of rare, threatened, and endangered species. It may provide advice to management and other agencies with respect to wildlife and fisheries management and research, to mitigation and compensation resulting from damage to wildlife habitat, and to wildlife education.

Comanagement boards are not instruments of self-government or self-management, because they are means of power sharing. They are instruments of public government for which the responsible government fish and wildlife agencies retain ultimate authority. The ministers of these agencies can adopt, reject, or vary the recommendations of the Board and also nominate the government representatives on these Boards. In practice, however, Board decisions are seldom overridden if the Boards establish their competence, credibility, and effectiveness among the parties.

The actual structure and powers of the boards under each claim vary somewhat. For example, the IFA provides for separate boards for wildlife and fisheries (charged with preparing management plans), and in the JBNQA the equivalent body is characterized as a coordinating committee. In the case of the IFA, Inuvialuit appointments to all joint management bodies are made by the Inuvialuit Game Council (IGC), an entirely Inuvialuit entity drawn from the Hunters and Trappers Committees (HTCs) of each community. The IGC and HTCs also provide advice to the joint management bodies. The HTCs may pass by-laws, subject in certain respects to the laws of general application, but the enforcement of all laws and regulations remains with the Crown, as is the case in all of the other agreements.

Comanagement, under the claims agreements, is not limited to fish and wildlife. Each agreement provides for a land use planning agency and some (NFA, YUFA) for separate water boards. The Gwich'in and Sahtu agreements provide for surface rights boards. These boards are advisory and do not replace existing government agencies. Members are appointed equally by governments and beneficiary organizations, except in the case of YUFA, in which the ratio is

2:1. Again, the scope and powers of these boards appear greater in the territorial final agreements than in the JBNQA, undoubtedly because in the latter the key development was already underway and the developer was a signatory to the agreement.

The comanagement system continues to require much technical and professional knowledge and research. However, the boards do not actually conduct any of this research (except in some cases for harvest statistics); they only advise governments in this regard. Both the technical paradigm and the database required to make it operational are the responsibility of the existing state management agencies. While it is the intent of most agreements to incorporate and utilize aboriginal knowledge (Chapter 15) and management systems (Chapters 11, 12, 16, and 17), none of the agreements specify how this should be done or what the criteria or tests of implementation might be.

Comanagement in practice

Experience across a broad spectrum of arrangements, both claims-based and ad hoc, indicates that there are several critical aspects of comanagement arrangements, at least from an aboriginal perspective.

Scope of comanagement

If one species is under a management agreement, but another species and (perhaps more importantly) its habitat are not, then comanagement will be of limited effectiveness. Those agencies responsible for habitat or for other resources in a multiple use situation are under no obligation to respond to the recommendations of the comanagement board. The problem is especially compounded if implementing these recommendations either costs money or involves foregone benefits. The problem is less significant in isolated northern areas where there are few conflicting uses. However, where there are many uses, all of which are tightly regulated by the Crown (as is more generally the case in southern Canada), a board with a narrow mandate may seem of limited value to its members.

Role of third parties

What is the role of other resource users: consumptive or nonconsumptive? One reason for the widely reported success of the BQCMB is that the management objective was never in dispute; the herds were to be managed for subsistence harvests by aboriginal residents of the range communities. Other potential users, however, are not directly represented on the board. In other cases, resolving conflicting management objectives are central tasks of the comanagement board. How stakeholders are identified and represented in the management system is obviously crucial. The view of government representatives that they also represent major non-aboriginal stakeholders, such as foresters, anglers, and hunters, is

Native Land Claims and Settlements in Canada

Several complex land claims have been submitted to the Government of Canada by aboriginal groups in the Yukon Territory, Northwest Territories (NWT), British Columbia, Quebec, and Labrador under a dual premise that (1) aboriginal title, where not extinguished by treaty or other governmental act, exists as a continuing legal right to lands and resources within traditional tribal territories, and (2) the Canadian federal government has not adequately discharged its obligations under an Imperial royal proclamation of 1763.

Increasingly, Native sovereignty and rights have been brought before the public and federal government since the early 1970s with many claims for lands and natural resources by various indigenous peoples of Canada. In 1973, the Government of Canada developed a comprehensive claims policy that expressed a willingness to negotiate land claims with organizations representing Natives. The major features of this policy are that aboriginal rights, titles, and interests to lands inside and outside a settlement area can potentially be exchanged for (1) surface and subsurface title to lands, (2) rights to the harvest of natural resources, (3) resource and environmental management rights, and (4) cash settlements. Claims have been settled with the Inuvialuit, Inuit, Gwich'in, and Sahtu in the NWT; with the Cree and Inuit of the James Bay region of northern Quebec; and with Native peoples of the Yukon.

In 1993, an agreement was signed for a large and complex claim between the Tungavik Federation of Nunavut and the Government of Canada. This agreement addresses Native land claims and harvesting rights within the region and commits the federal government to create a separate territory of Nunavut, whose government will be under effective Inuit control with a population that is 80% Inuit. Nunavut will comprise 200 million hectares of northern Canada, including Crown lands on which Inuit can harvest natural resources. Inuit will comanage land-use planning, wildlife, environmental protection, and offshore resources of Nunavut in cooperation with the federal government and other entities where appropriate. Comanagement is also a key element of other agreements between Native peoples and the federal government (see Chapter 13).

Nunavut is scheduled to be formed on April 1, 1999. This will provide Inuit with more autonomy than any other Native group in Canada. Many other claims are still in progress, and the complex process of developing settlements will continue into the foreseeable future. The settlement of claims in Canada is an important step for Native peoples and will have profound implications for the management of natural resources. It also has possible global significance for other native populations in Australia, Africa, and South America, who could use Native land claims in Canada as a model for pursuing their own claims.

Christy M. Parker

seldom accepted by those groups, and this has become a major stumbling block in negotiating land claims in southern Canada.

Mandate and accountability

Comanagement boards are formal instruments of public government, and members are therefore ultimately accountable to the minister of the Crown to whom the board reports. In the case of the BQCMB, for example, all members are ministerial appointments, although in practice, user members are named by the political bodies that represent users, and such designations are never refused. In practice, members operate as representatives of the bodies that named them. One result is that government members have a much clearer sense of their mandate and represent the interests and objectives of their agency, within which formal positions are likely to have already been taken on matters coming before the board. Aboriginal members are less likely to have such a clear and formal mandate on specific issues, and one result is that the government agenda often prevails, with aboriginal members assenting, both because of restricted mandates and because some important underlying issues do not reach the table. In contrast, under the IFA, aboriginal members are accountable to the Inuvialuit Game Council, a policy-setting body with a clearly developed agenda (Bailey et al. 1995). Some boards are formally assigned the responsibility for developing a management plan that will be implemented primarily by the appropriate Crown agency. Others have primarily a review function and recommend for or against actions proposed by others.

Communication

A good board with low member turnover and regular attendance can develop as a team; mutual respect and understanding can help overcome longstanding differences at the board level. But this can have only limited impact if the wider public does not understand and agree with the board's decisions. Effective communication among the scattered, isolated communities of the North is difficult and expensive, and the resources for it are scarce. Furthermore, the traditions of decision making and implementation can vary substantially between government agencies and aboriginal communities. Both board meetings and board secretariats (where they exist) operate more in the government than the aboriginal style.

Research

Comanagement boards at best supplement, but do not replace, existing resource management agencies. Most have either no secretariats or purely administrative ones. Accordingly, boards normally get the technical advice required for planning and decision making from resource management agency scientists, but this is not neutral information. Some advocate that aboriginal or user members of boards obtain independent technical advice, but whether they can actually do so depends on funding levels and operating procedures. Only some of the claims-based boards have been successful in doing this, with the most outstanding examples being the various comanagement boards established under the IFA.

The definitions of conservation found in the agreements assume particular importance. Terms contained in these definitions, such as populations, productivity, and ecological systems, would appear to require the type and quality of data that wildlife managers traditionally use, without an appropriate reference to the knowledge structure of the aboriginal management system. However, the direct linkage of conservation to aboriginal harvesting rights, as in the IFA, has resulted in significant aboriginal control over research priorities and design.

Use of traditional knowledge

Aboriginal self-management systems are based in what is often referred to as traditional knowledge (Chapter 15), which in turn is incorporated in language. How well do comanagement systems account for and incorporate these?

The wildlife comanagement regimes raise significant problems of definition and terminology. All of the final agreements have wildlife sections (and definitions of wildlife), although that term is not an objective description but a cultural statement of the relationship of people and animals (and habitat) in an agricultural, settler heritage. It appears to have no direct equivalent in aboriginal languages. Likewise, conservation has been defined exclusively in the technical idiom of biology and of Western paradigms of resource management. This suggests the power of the dominant system, through language alone, to subtly but inexorably force aboriginal organizations into negotiating on unfamiliar turf. This is done not simply in technical jargon but in a non-indigenous paradigm and knowledge system relating to all aspects of the use and management of natural resources.

A further problem is how traditional knowledge is incorporated into comanagement. Despite nominal commitments, both the structures and the idiom almost invariably require English-speaking participants and use non-aboriginal concepts and paradigms to actuate the management system. For example, wildlife managers often explain scarcity and abundance differently from how traditional harvesters explain them, although the boards are expected to produce consensus on total allowable harvests. The everyday tools of wildlife managers, such as stock assessment, herd definition, surplus, and sustainable yield, are not necessarily shared as key concepts by traditional harvesters. There is not necessarily a shared understanding of what management is, or of what its objectives and tasks might be (e.g., writing management plans).

The key terms and concepts of the wildlife provisions can be quite ambiguous, especially from a cross-cultural perspective, and are thus properly the central terrain of both negotiations and implementation. This may be part of the reason that no claimant group has been as successful in negotiating its opening position on management rights as it has on harvesting rights. However, continuing to ignore this problem by supposing that the science of wildlife management is universal and value-neutral will neither satisfy the participants nor prevent resource depletion or harvest disruption.

Scarcity

Failure to incorporate traditional knowledge is a problem, because a key test of a board, in the eyes of both the aboriginal and non-aboriginal public, is how it deals with scarcity. The BQCMB was born of a perceived scarcity crisis, but almost immediately after it came into being it became apparent that the herds were more numerous than supposed and that the drastic measures that some had proposed were unnecessary. Instead of being an emergency response team responding to crisis, the board had some breathing room in which to develop a cooperative atmosphere and a management plan. The continued health of the herds cannot, however, be directly attributed to the management skills of the Board or its supporting agencies, except insofar as the improved sense of security among hunters, and of understanding between hunters and managers, may have improved the climate for self-regulation.

The effect of this good fortune, however, is that the board has never been tested by scarcity, and that may still be its most crucial future test. What the board can undoubtedly claim as a success is in contributing to, if not indeed creating, an atmosphere of mutual recognition, tolerance, and understanding that will be essential for dealing with any future crisis (Usher 1993). Negotiated harvest reductions in response to local scarcity have occurred in the western Arctic under the IFA comanagement system (Bailey et al. 1995).

Formalizing informal management systems

For aboriginal participants, the establishment of comanagement rather than self-management structures formalizes a previously informal system. Several responses are possible:

1. Direct representation by elders and traditional harvesters. This requires the system (including the managers and scientists) to accommodate fully to the use of aboriginal languages, knowledge, and procedures.
2. User representation, also involving experienced people but ones who can accommodate themselves to the state system, even if the cost is more passive participation on their part.
3. Training of young aboriginals to participate in the dominant technical idiom (i.e., to become, in the state's terms, qualified biologists and resource managers who could sit on comanagement bodies without technical assistance or provide technical assistance to board members).
4. Assignment of non-aboriginal technical or legal advisors to participate in these technical committees, with aboriginal harvesters or politicians providing at best inconsistent and occasional attendance.

It would appear that the second and fourth strategies are the most common ones. The IFA provides the closest approximation of the first, but perhaps this is because English is widely spoken in the western Arctic and the cultural gap between resource management concepts may not be so great as in other areas. Examples of the third strategy are also rare. Despite a substantial increase in Native Northerners obtaining post-secondary education, very few have opted for advanced study in the natural sciences and resource management.

Nonetheless, comanagement does at a minimum result in an open discussion of research and management techniques that formerly occurred behind closed doors. Although research and management rarely incorporate aboriginal knowledge and concepts, managers do have to justify and explain what they are doing (Chapters 16 and 21), and in some cases will not undertake certain programs to which harvesters clearly object.

Some outstanding issues in comanagement

Some aboriginal groups are satisfied with existing comanagement arrangements. They find that they not only suit their needs well, but that they can use this system to their advantage. Others find them at least acceptable, because they are a significant improvement over the former closed-door system of management. Still others have no desire to comanage resources with outsiders, but seek exclusive management authority within a limited geographical area. There is some indication that comanagement is more likely to be preferred if migratory or transboundary populations are involved, because such regimes bring both governments and users together among jurisdictions. Perhaps not surprisingly, those groups most dependent on migratory species, such as caribou (*Rangifer tarandus*), waterfowl, and marine mammals, are the strongest advocates of comanagement and regard it as the key to resource conservation and social and political stability, despite its day-to-day problems and frustrations.

It is not clear where the majority of harvesters stand. Perhaps they are reserving judgment, knowing as they have for generations that the ultimate decisions about use and conservation belong to those on the land, not those in the office. It is important to remember that comanagement represents, for them, a compromise. It is much less than self-determination or self-regulation, which is what many aboriginal harvesters actually want.

Will comanagement regimes protect the mixed, subsistence-based economy that characterizes most northern aboriginal communities? If a fundamental dimension of aboriginal title and rights is a coherent and viable indigenous management system, then it seems appropriate to ask if comanagement regimes will promote and protect the indigenous system or be a means of incorporating it into the state system.

What will administration and decision making look like when there are numerous and to some extent overlapping comanagement arrangements, as seems possible in at least some jurisdictions, perhaps especially where they are not claims-based? Governments like "single windows", but these could become many and overlapping. At present, ministerial override is rare, and bureaucrats at least pay attention to the boards. But when a few become many, and their objectives and recommendations conflict, that situation is likely to change.

What does this tell us about dealing with climate change? Climate change will likely affect the range and habitat requirements of key economic species, their abundance, and the timing of their appearance (Chapters 4–8). It seems likely that the immediate effects of global change on resource harvesters and their communities will be adverse (Roots 1993). We can anticipate that change will be required in resource harvesting patterns and needs and in resource management. Dealing with this effectively will require both an understanding of the problem and preparedness to respond to it and implement changes. For those dependent on these resources, this implies some degree of control over the situation and an adequate level of trust between interdependent parties. Comanagement, understood as real power sharing, provides a basis for this.

The information base that provides understanding of the problem and its solutions, and the public acceptance of it, will also be critical. Harvesters and resource-dependent communities must be convinced that the "expert" view on climate change and what to do about it converge with their understanding of the situation on the ground and with the cultural and historical context in which they integrate and interpret this experience. Comanagement may be a useful venue for this dialogue, but that depends on the successful use of traditional knowledge and perspectives and a better integration of state and indigenous management systems. The jury is still out.

References

Bailey, J. L., N. B. Snow, A. Carpenter, and L. Carpenter. 1995. Cooperative wildlife management under the Western Arctic Inuvialuit land claim. In *Integrating people and wildlife for a sustainable future. Proceedings of the first international wildlife management congress*, edited by J. A. Bissonette and P. R. Krausman, 11–15. Bethesda, MD: The Wildlife Society.

Osherenko, G. 1988. *Sharing power with native users: Co-management regimes for Arctic wildlife*. CARC Policy Paper 5. Ottawa: Canadian Arctic Resources Committee.

Roots, E. F. 1993. Climate change—Its possible impact on the environment and the people of northern regions. In *Impacts of climate change on resource management in the North*, edited by G. Wall, 127–151. Department of Geography Occasional Paper 16. Waterloo, Ontario: University of Waterloo.

Usher, P. J. 1987. Indigenous management systems and the conservation of wildlife in the Canadian North. *Alternatives* 14: 3–9.

Usher, P. J. 1993. The Beverly-Kaminuriak Caribou Management Board: An experience in comanagement. In *Traditional ecological knowledge, concepts and cases*, edited by J. T. Inglis, 111–120. Ottawa: International Program on Traditional Ecological Knowledge and International Development Research Centre.

14

Common property resource management and northern protected areas

John Wiener

There are four features of traditional common property resource management that are relevant for the management of protected areas, local peoples, and the challenge of climate change. First, traditional resource management systems were developed through long-term experience with local conditions and resource responses (Driver 1969; Damas 1984), resulting in a detailed understanding of local resource conditions and relationships (Chapters 9, 13, and 15). Second, the resilience of traditional common property management is important for uncertain conditions regarding climate. Resilience is the capacity of a system to maintain its elements in similar but varying relationships despite perturbations (Clark and Munn 1986).

Third, fundamental economic choices in common property regimes are different. Traditional common property resource management within living cultures does not operate with a strong preference for the present and a high discount rate for the future. Indigenous peoples have ethics that do not treat the future as worth less than the present or accept some amount of fish as an adequate substitute for deer. In contrast, economists tend to presume mobility or transformation of capital, meaning that resources can always be substituted.

Fourth, compared to industrial societies, the resource demands of people living in common property resource conditions (and certainly many northern peoples) are minimal. Northern indigenous people have until recently been largely self-sufficient, although trade, newcomers and emigrants, occasional hostilities, and other outside contact have also been important (Fienup-Riordan 1994). Traditionally, people assumed roles defined by age, sex, and their abilities, within a fabric of social rights and obligations. The behavioral adaptations that developed over millennia, under challenging conditions, include resource management regimes and the inseparable cultural values and norms that supported these regimes.

Why do we need better management?

A profound cultural revolution accompanied by environmental, political, and economic turmoil, is occurring around the world. Debates about the relationship between individuals and society, the role and limits of sovereign states, conventional political and economic theories, and the limits of science challenge the dominant views of the twentieth century. These debates affect management institutions everywhere. The notion that there is one correct answer to any question is eroding. In its place is the increasing acceptance of subjective answers that are relative to specific cultural and natural contexts. There is less interest in top-down, expert control and more interest in bottom-up, local management, including local participation in planning and implementation.

Arctic and subarctic indigenous cultures around the world are achieving increasing political power, skill, and scientific capacity (Osherenko and Young 1989). They are increasingly exposed to southern material wealth and involved in commercial economies, with a range of adaptations from remote villages to oil boom cities (Chance 1990; Jorgensen 1990). Industrial activities now threaten some environments, as do immigration and visitation by Southerners. Nonetheless, natural resources remain important for material as well as cultural purposes (Usher 1986; Berkes et al. 1994). Northern indigenous cultures are gaining increasing political power, and they have an undeniably large stake in the management of northern natural resources.

Although the application of the concept of poverty to non-Europeans voluntarily pursuing traditional lifestyles is problematic, it is clear that northern indigenous people experience considerable relative deprivation and health problems (Chapter 3).[1] Although many mixed cash-subsistence economies provide a reasonable standard of nutrition (Chapters 10 and 11), reliance on imported goods creates vulnerability and a spiral of dependence on these goods.[2] Lack of local control over resources is common to current theories of rural poverty and is clearly relevant for protected areas (Fortmann 1990; West and Brechin 1991). The old ways seem futile, the new ways are unavailable, and collective abilities to control resources are eroded. Traditional skills and values seem irrelevant to youngsters who fear that their heritage is obsolete. This is an old story, not confined to the North, but when combined with the widely perceived view that central government management is either too expensive or inefficient, it provides motivation to consider local strategies of resource management that were traditionally viable.

---

[1] The natural resource shortage component of Northern poverty is minimal. In Alaska, for example, the subsistence portion of total wildlife harvest is about 4%.

[2] For example, pressures for centralized settlement and access to services (schools, health care, cash opportunities) are balanced against maintenance of trap-lines and hunting territories. People increasingly rely on modern technology, especially motorized transport, for efficient use of time and greater access to resources (Langdon 1986; Brody 1987; Wolfe and Walker 1987).

Theory of common property regimes

Common property systems of management require reconceptualization of the dichotomy between private and state property and may also involve a related notion of comanagement.

Private property and state property

Property is a social construction: a relevant group (from village to nation-state) respects and defends claims to a stream of benefits from a resource, which can be a physical object, a social obligation, or anything else (Bromley 1991). The rights are power to control, but they are almost never unconditional. Individuals in current legal systems in Canada and the United States have rights that are limited in terms of impositions on others, and by the capacity of the state to reassign the property rights, sometimes with compensation. Rights are freely assignable for time periods from short to forever and to almost any entity. Some societies may assign more limited rights to individuals, families, clans, and user groups. Property rights can be described in seasons (common in North America), states of the resource (age of animals, ripeness), or uses (grazing or firewood use; food harvest but not harvest for commercial sale). Rights can be assigned on the basis of purchase, licenses, kinship, or any other basis. Fortmann and Bruce (1988) discuss tree property rights, and McCay and Acheson (1987) and Bromley (1992) present examples from around the world. Some groups do not assign some rights at all (e.g., the right to destroy certain resources, rights to own human beings).

By conceptualizing property rights as a continuum rather than a dichotomy, the following can be distinguished: (1) no one claiming rights ("non-property," open access), (2) a single individual claiming ownership (private property), (3) a definable user group claiming joint ownership (common property), or (4) a sovereign claim of state property. Almost no property right is unconditional, unlimited, and unaffected by the collective will that created it. Even the right to one's own life can be lost by social decision (Bromley 1991; Fienup-Riordan 1994).

Comanagement versus property rights

Because rights to a stream of benefits can be described and divided at will, there is no inherent difference between rights of ownership and rights of disposal or other rights. Comanagement, the shared authority between a government and a local group over the management and allocation of resources (Osherenko 1988), is widespread in the North (Caulfield 1992; Pinkerton 1992; Dyer 1994; Chapter 13, this volume). Important issues include decisions about who can harvest, quantities harvested, policy enforcement, payment of necessary costs, and allocations of rights to harvest (Bromley 1992).

Conceptually, there is no difficulty in allocating rights; the difficulties are in overcoming the resistance of those who perceive loss (Pinkerton 1992). Those who control the state, however, may resist delegation of state powers because the delegates may not be as easily controlled. In Alaska, for example, commercial and non-Native sport hunting and fishing interests dominate the state agencies, excluding Natives from power over resources (Noble 1987; Case 1989). Exclusion through instrumental use of the state is quite familiar in the North (Langdon 1986; Marchak et al. 1987; Pinkerton 1989) and the rest of the world.

The theory

Within the intellectual context and material success of the West, it is understandable how "hard science" promoted economic analysis centered on mathematical formalizations and analytical concepts such as private property. Economists seeking rigor adopted their model because the results were theoretically powerful and practically useful.

If common property regimes—in which a user group managed and allocated a known resource for its own purposes—were as ubiquitous as we now know, why were they overlooked for so long? There are several explanations. First, there is a strong link between the dominant capitalist powers of the modern age, and the ideology that legitimates the distribution of power and rights. Second, in the era of modernization (after the decline of formal colonialism), there were compelling arguments for supplying the new sovereignties with the ideology and structures that had apparently been successful for the West. Third, recommending communal ownership might arouse suspicion, given the anti-Communist feelings through the Korean War, Vietnam War, and Cold War periods. Capitalism, using maximum private property and some state property, seemed to provide wealth (negative outcomes tended to be overlooked). Communism, using minimal private property and maximum state property, seemed to provide only misery and economic failure (positive accomplishments tended to be overlooked). In a world caught up in an apparent struggle between private property versus state property, it is understandable that a middle ground was rarely considered.

The concept of open-access resources was first formalized by Gordon (1954) and Scott (1955). Referring to nonproperty, or open-access fisheries (which they unfortunately called common property), they showed that users without ownership interests would lack incentives to manage a fishery for maximum economic yield, maximum biological yield, or sustainability. The commons problem, as they called it, stemmed from users' ability to take as much as they wanted and for all potential users to take the resource. Self-restraint merely increased supply for others, and dissipation of economic rent was inevitable. The result was that the resource would be overused and possibly ruined. Garrett Hardin generalized from this to *The Tragedy of the Commons* (1968), which was remarkably compelling in its logic. However, many of Hardin's premises about social phenomena were so unrealistic that they are rarely confirmed by observable cases (McCay and Acheson 1987; Bromley 1992).

Despite overwhelming self-interest, individuals often achieve better results from collective action. Even with some cheating, there are strong incentives for individuals to cooperate as group members. Optimal management is achievable by means other than individual private property, or state property with coercive enforcement (Bromley 1992). Villages or user groups can and do function as firms, with the locus of economic agency shifted from individuals to the group, with or without ideological support in cultural coding or ethical teachings. The village operates as a firm, with appropriate divisions of labor. Men travel dangerous terrain or open water and hunt large animals, while women supervise child care, foraging, and food processing. Elderly persons act as consultants, management experts, and libraries. Reciprocity through the life-cycle and between families, as well as between villages, is the basic tool of risk management. Common property can be economically optimal, depending on the nature of the resource, its use, and the users.

Common property may best serve the operation of a resource-using system for long time periods. If the economic agent is a group, it may be less susceptible to time preferences that result in strongly discounting the future and can make it economically rational to extirpate resources or even species (Clark 1989). Time and security of interests are central to any given incentive structure whether it is the open-access described by Hardin or a comanagement regime or other assignment of rights. A private owner would consider the consequences of overharvesting, while non-owners would not bear the costs of abuse, so ownership was clearly the ideal state of resource management (Demsetz 1967; Furubotn and Pejovich 1972). This fits the capitalism versus communism ideology and supported radical redistributions of property into private hands all over the Third World, from the local scale through governments. Because only private ownership would maximize productivity and the well-being of society, it is assumed that resources should be privatized. This argument is still popular, especially when coupled with current anti-government sentiment. Within economics, however, confusion over the term commons is finally being reduced, with the acceptance of distinction from open-access resources.

The idea of a regime

The term *regime* has become popular for denoting a set of institutions that control the management of a resource (Young 1982). A regime denotes a resource and the rules of its use. Some resources certainly lend themselves to some kinds of rules, such as those benefits from which no one can be excluded ("public goods," such as national defense or immunization), and some resources lend themselves to individual property (such as my lunch). Optimal allocation of rights is determined by the nature of the resource and the desired results. Common property regimes are a likely alternative if the benefits of private rights are not worth the maintenance costs. A resource management regime includes the formal rules and agencies of implementation, as well as the necessary cultural underpinnings that motivate people to cooperate and make the regime successful. Regimes fail when they lack sufficient support to achieve their purpose, such as when they are not adequately implemented. Regimes also fail when local people refuse to comply, or outsiders cannot be made to comply. Some fisheries provide examples of failure to preserve fish and local fishermen, especially for situations in which locals have no control of resources harvested at unsustainable rates by industrial fishing operations (McGoodwin 1990). From the Native point of view, this is a common scenario in Alaska and the Pacific Northwest (Langdon 1986; Marchak et al. 1987; Pinkerton 1989).

Security of interests

A defined user group can manage and sustain resources efficiently without individual or state ownership and can provide benefits sufficient to reward cooperation. These benefits include: (1) defense of relevant rights by the appropriate level of organization, and (2) security of tenure in those rights over sufficient time to acquire the benefits of present restraint and investment. Traditional regimes of many Native peoples in the North (Berger 1985; Nelson 1986; Brody 1987; Chance 1990) include a fabric of socially honored rights and obligations in wildlife, in the fruits of others and in performance of social roles (Usher and Bankes 1986). Risk management and resource allocation evolved over long periods of time, and individuals accepted their identities and roles, including acceptance of restraints on overharvest and obligations to others. A privatized formal system of rights would have been inefficient. The costs of information on resources and others' activities, contracting for cooperation and division of labor, and policing arrangements and behavior would have been effectively subsumed into cultural learning about what a person knew and did as a member of the group. The well-being of the individual was inseparable from the well-being of the group and of the harvest on which the group depended.

In the North, individualized property rights may have been created in traplines in response to the appearance of a market for furs that were traded by individuals in most cases, but these were not rights to other resources (Brightman 1987). One of the most important aspects of the fur trade was the individualization of resource exchange between persons and the trading company. Individualized economic interactions varied in importance, until the widespread availability of imported tools and equipment, cash transfers, and wages. Despite economic modernization, subsistence sharing and reciprocity within groups is still highly valued spiritually and materially (Brody 1987; Berkes et al. 1994; Chapter 10, this volume).

In the past, group resources were defended as necessary (Driver 1969; Damas 1984; Fienup-Riordan 1994) and were literally managed as assets (Brody 1987). Abuses were difficult to hide and were sanctioned by disapproval (from childhood teaching, violence, ostracism). National governments, however, deny the right to secede and assert authority over both resources and Natives. Governmental intervention has been justified as part of modernization and promotion of economic efficiency, although such claims are made only by those who benefit from the action. There is an urgent need for restoration of secure rights in resources to the culturally appropriate holders.

Insecurity and inappropriate incentives

Tragedy can be created by governmental regimes that cannot be effectively implemented, but that can corrupt officials, criminalize previously acceptable activities, and destroy security of access to resources. In Nepal, forests were nationalized in an effort to protect them from local users after many centuries of sustained use by villagers working within known rules and a few years of increased use by tourists. Normal and necessary uses were criminalized, and incentives to maintain the resource were destroyed; it was only by restoration of secure rights to villages that the forests were saved (Blaikie and Brookfield 1987; Bromley 1992). If Northerners are denied legal access to normal harvests, they might be expected to respond as the Nepalese did. Without secure rights to wildlife, poaching could become accepted behavior.

Scarcity caused by closure of protected areas or climate change could lead to disregard of the law and severe risk for scarce resources. Being at risk for resources and acting illegally may well be associated with acting outside traditional law and culture. But how would one expect a rational person to behave if he/she has no right to an open access resource? I predict that walrus (*Odobenus rosmarus*) "head-hunting" for ivory for illegal Asian markets will be the first serious scandal, unless the Alaska Eskimo Walrus Commission is successful in urging restraint by non-Natives as well as Natives (Langdon 1986; Caulfield 1992; Fay et al. 1994).

Maximum economic yield and resilience

Once a common property user group is acknowledged as an economic agent, there are two important consequences in traditional northern resource management. First, the time dimension changes; the village persists and survives by not considering the present as more valuable than the future. A moose (*Alces alces*) in the future matters, so careful management of moose in the present is only common sense. Economic thinking has no conflict with other values. Second, the agent will be rewarded for management to achieve maximum economic yield, rather than maximum sustained yield.

Maximum economic yield (mey) is attained by harvesting the level at which the difference between cost of harvesting and benefit of harvesting is greatest: the best possible return on investment (Figure 14.1). The maximum sustained yield (msy) is biologically defined, and harvest at that level produces the largest num-

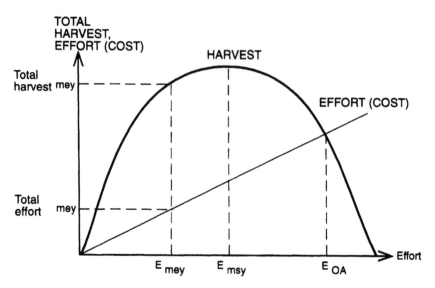

Figure 14.1. *The relationship between maximum economic yield (mey), maximum sustained yield (msy), and effort (E) required to attain them. Effort that would be exerted under conditions of open access (no property rights in effect) is also shown (E_{OA}). Note that harvest levels differ dramatically and that economic rents are zero in OA and maximized at mey levels. From Townsend and Wilson (1987), p. 317, figure 15.3. Reprinted with permission of University of Arizona Press.*

ber of fish or other resource item. The best economic yield comes from adjusting the number of harvesters and harvested resources to the point at which *mey* is comfortable for the harvesters; they can use other capacity for other activities. The *mey* harvest level is below the *msy* level. This is critical because the difference is a cushion. In a bad year for the resource, the harvesters will more likely still be taking stock within the stock's capacity to be replaced. In a bad year for the harvesters, the extra effort needed to harvest more resources is still rewarded (costs are less than benefits, and the activity is still successful and profitable), and that additional level of harvest does not necessarily threaten the resource.

There may also be cases in which the marginal cost of harvesting too much is minimal after harvesting just right; fish traps might be such a case. Adjusting harvest would require a careful decision to stop at the right point, involving cultural learning and values. When the marginal cost of harvesting declines, cultural and conscious factors might be needed to limit harvest as well. Ethics against waste are important. This may apply to harvests of herd animals such as migratory caribou (*Rangifer tarandus*). Dispersed animals generally maintain territories of their own, so harvest effort (cost) increases with each added animal, and the cost line will continue to rise (increasing marginal costs).

Treating the user group as a single economic agent, like a firm, makes it rational for the group to act in the way property right theory suggests is ideal for individual owners. However, this owner is not bound by the high discount rates of Western society unless security of interests is threatened. Without security, the incentives for conservative management are gone. Without a vital collective life, the benefits of group management are gone. Wildlife managers have a vital interest in the cultural life of local people, who are an important part of the ecosystems that they inhabit.

Summary

What can we expect in northern protected areas under conditions of local poverty, threats to security, and the stresses of climate change? The most reasonable approach is to develop comanagement arrangements that provide local residents with a long-term right to a desirable harvest level of wildlife; this allows access to the power of the state to defend the environmental basis of that harvest. Governments may not be willing to transfer the full range of property rights, but this is not necessary to achieve free (to the state) use by local residents, who are essentially expert game wardens and resource managers with long records of success in resource protection.

Maintenance of traditional cultures with security of rights to wildlife in protected areas—an intentional reversion to historic patterns—is the most progressive approach for creating self-sustaining systems. This approach will preserve both natural and cultural values in northern regions facing potential stress from environmental change. Common property regimes are efficient and sustainable social structures for wildlife management. However, no model is universally applicable, and local conditions and human needs vary. The richness of adaptation demands local participation in all phases of management. The role of the state, above all else, is to defend these systems. The most promising direction is to work with the underlying ideas common to comanagement (Chapter 13), leaseholds, group licenses, and harvest regulation. The design of these arrangements will never be easier than it is now, when we know how important it is to "save all the pieces."

References

Berger, T. R. 1985. *Village journey*. New York: Hill and Wang.

Berkes, F., P. J. George, R. J. Preston, A. Hughes, J. Turner, and B. D. Cummins. 1994. Wildlife harvesting and sustainable regional Native economy in the Hudson and James Bay Lowland, Ontario. *Arctic* 47:350–360.

Blaikie, P., and H. Brookfield. 1987. *Land degradation and society*. London: Methuen.

Brightman, R. A. 1987. Conservation and resource depletion: The case of the boreal forest Algonquins. In *The question of the commons*, edited by B. J. McCay and J. A. Acheson, 121–141. Tucson: University of Arizona Press.

Brody, H. 1987. *Living Arctic: Hunters of the Canadian North*. Vancouver: Douglas and McIntyre.

Bromley, D. W. 1991. *Environment and economy: Property rights and public policy*. Oxford: Blackwell.

Bromley, D. W., ed. 1992. *Making the commons work*. San Francisco: Institute of Contemporary Studies.

Case, D. S. 1989. Subsistence and self-determination: Can Alaska Natives have a more "effective voice?" *University of Colorado Law Review* 1989:1009–1035.

Caulfield, R. A. 1992. Alaska's subsistence management regimes. *Polar Record* 28:23–32.

Chance, N. 1990. *The Inupiat and Arctic Alaska: An ethnography of development*. Fort Worth, TX: Holt, Rinehart, and Winston.

Clark, C. W. 1989. Clear-cut economies. *The Sciences* 29:17–19.

Clark, W. C., and R. E. Munn, eds. 1986. *Sustainable development of the biosphere*. Cambridge: Cambridge University Press and International Institute for Applied Systems Analysis.

Damas, D., ed. 1984. *Handbook of North American Indians, Vol. 6—Arctic*. Washington, DC: Smithsonian Institution.

Demsetz, H. 1967. Toward a theory of property rights. *American Economic Review* 57:347–359.

Driver, H. E. 1969. *Indians of North America*. 2d ed. Chicago: University of Chicago Press.

Dyer, C. L. 1994. Proaction versus reaction: Integrating applied anthropology into fishery management. *Human Organization* 53:83–88.

Fay, F. H., J. J. Burns, S. W. Stoker, and J. S. Grundy. 1994. The struck-and-lost factor in Alaskan walrus harvests, 1952–1972. *Arctic* 47:368–373.

Fienup-Riordan, A. 1994. Eskimo war and peace. In *Anthropology of the North Pacific Rim*, edited by W. W. Fitzhugh and V. Chaussonnet, 321–335. Washington, DC: Smithsonian Institution.

Fortmann, L. 1990. Locality and custom: Non-aboriginal claims to a customary usufructuary right as a source of rural protest. *Journal of Rural Studies* 6:195–208.

Fortmann, L., and J. W. Bruce, eds. 1988. *Whose trees? Proprietary dimensions of forestry*. Boulder: Westview Press.

Furubotn, E. G., and S. Pejovich. 1972. Property rights and economic theory: A survey of recent literature. *Journal of Economic Literature* 10:1137–1162.

Gordon, H. S. 1954. The economics of a common property resource. *Journal of Political Economy* 62:124–142.

Hardin, G. 1968. The tragedy of the commons. *Science* 162:1234–1248.

Jorgensen, J. G. 1990. *Oil age Eskimos*. Berkeley: University of California Press.

Langdon, S. J., ed. 1986. *Contemporary Alaskan Native economies*. Lanham, MD: University Press of America.

Marchak, P., N. Guppy, and J. MacMullan, eds. 1987. *Uncommon property: The fishing and fish-processing industries in British Columbia*. Toronto: Methuen.

McCay, B. J., and J. M. Acheson, eds. 1987. *The question of the commons*. Tucson: University of Arizona Press.

McGoodwin, R. 1990. *Crisis in the world's fisheries.* Stanford, CA: Stanford University Press.

Nelson, R. K. 1986. *Hunters of the Northern forest.* 2d ed. Chicago: University of Chicago Press.

Noble, H. 1987. Tribal powers to regulate hunting in Alaska. *Alaska Law Review* 4:223–275.

Osherenko, G. 1988. Can comanagement save Arctic wildlife? *Environment* 30:6–13, 29–34.

Osherenko, G., and O. R. Young. 1989. *The age of the Arctic: Hot conflicts and cold realities.* Cambridge: Cambridge University Press.

Pinkerton, E., ed. 1989. *Cooperative management of local fisheries.* Vancouver: University of British Columbia Press.

Pinkerton, E. W. 1992. Translating legal rights into management practice: Overcoming barriers to the exercise of comanagement. *Human Organization* 51:330–341.

Scott, A. D. 1955. The fishery: The objectives of sole ownership. *Journal of Political Economy* 63:116–124.

Townsend, R., and J. A. Wilson. 1987. An economic view of the tragedy of the commons. In *The question of the commons*, edited by B. J. McCay and J. M. Acheson, 311–326. Tucson: University of Arizona Press.

Usher, P. J. 1986. *The devolution of wildlife management and the prospects for wildlife conservation in the Northwest Territories.* CARC Policy Paper 3. Ottawa: Canadian Arctic Resources Committee.

Usher, P. J., and N. D. Bankes. 1986. *Property: The basis of Inuit hunting rights— A new approach.* Ottawa: Inuit Committee on National Issues.

West, P. C., and S. Brechin, eds. 1991. *Resident people and national parks: Social dilemmas and strategies in international conservation.* Tucson: University of Arizona Press.

Wolfe, R. J., and R. J. Walker. 1987. Subsistence economies in Alaska—Productivity, geography, and development impacts. *Arctic Anthropology* 24:56–81.

Young, O. R. 1982. *Resource regimes, natural resources and social institutions.* Berkeley: University of California Press.

15

Inuit indigenous knowledge and science in the Arctic

Ellen Bielawski

. . . the process of opening Western knowledge to traditional rationalities has hardly yet begun. (Salmond 1985, p. 260)

I am telling you about myself. You didn't even bother telling me about yourself, you just wanted me to write stories about myself. I don't think that's fair. I would like to know about your parents and I would like to know about other things. I am an old man now and I am curious. (Akuliaq 1967)

. . . the vast and particular knowledge of the Eskimo, garnered from hundreds of years of their patient interrogation of the landscape, was starting to slip away. (Lopez 1986, p. 6)

Two groups of people—informants—are currently enriching my knowledge of the Arctic. One group is made up of Inuit elders, who speak little, if any, English. The other is scientists who work in the Arctic.

My colleagues have commented on how different these two groups are. Inuit elders, born on the land, spend most of their lives as subsistence hunters; scientists in their pursuit of pure knowledge are as passionate as Inuit hunters are about the rich taste of caribou (*Rangifer tarandus*) marrow. I respond lightly, "I work

This article appeared in a slightly different form in *Northern Perspectives* 20:5–8 (1992). Reprinted with permission of the author and Canadian Arctic Resources Committee. *Northern Perspectives* is the quarterly journal of the Canadian Arctic Resources Committee (CARC), a membership-based public interest organization with offices in Ottawa. For details about joining CARC, please write to: Canadian Arctic Resources Committee, Suite 412, 1 Nicholas Street, Ottawa, Ontario, Canada, K1N 7B7. Research for the theoretical context of the case studies was supported by a Killam Postdoctoral Scholarship at the University of Alberta. The Inuit Relocation case study was supported by Multiculturalism Canada, Department of the Secretary of State. The Ethnography of Arctic Science is supported by the Social Sciences and Humanities Research Council of Canada, Science, Technology and Policy Program. I am very grateful to the Documentation Center, Avataq Cultural Institute, for use of the Inuit Archive, and especially to Sylvie Côté Chew for her invaluable assistance there.

with elders who speak no English, and it's quite similar to working with scientists, because they don't speak any English either."

I am, of course, trained in the Western scientific tradition, and not fluent in Inuktitut (the language spoken by the Inuit). But I find much in common among the informants in the two case studies I have in progress. Both have difficulty communicating their knowledge to those who use it. Both are isolated from much of the knowledge held by the other. In recent history, they have seen little use for each other's knowledge.

The research I am doing is about the difference between Inuit indigenous knowledge and Arctic science. Conflict between ways of knowing began when Inuit (whose ancestors first occupied the Arctic nearly 4000 years ago) and European explorers first met. The conflict is sometimes subtle and quietly as well as savagely devastating to Inuit, who nevertheless endure. The conflict continues in the form of negotiations for land, sea, and resources; for political power; for housing and health care; for culture. The difference between Inuit and Western knowledge underlies conflict in all realms. The conflict is detrimental to both cultures.

Current thinking recommends that Arctic scientists and those who use their work (managers and policy makers) resolve the conflict by recognizing the continuing existence and value of indigenous knowledge (also called traditional, or local, knowledge). Canada's former Minister of the Environment has written, "Our task is to integrate traditional knowledge and science" (de Cotret 1991). The Traditional Knowledge Working Group, created by the Government of the Northwest Territories, strives for legislative and policy changes that will integrate traditional knowledge into policy about wildlife, health, justice, and social problems. The working group debated the meaning of traditional knowledge for more than two years before reporting:

> The lack of common understanding about the meaning of traditional knowledge is frustrating for those who advocate or attempt in practical ways to recognize and use traditional knowledge. For some, traditional knowledge is simply information which aboriginal peoples have about the land and animals with which they have a special relationship. But for aboriginal people, traditional knowledge is much more. One elder calls it "a common understanding of what life is about." Knowledge is the condition of knowing something with familiarity gained through experience or association. The traditional knowledge of northern aboriginal peoples has roots based firmly in the northern landscape and a land-based life experience of thousands of years. Traditional knowledge offers a view of the world, aspirations, and an avenue to "truth," different from those held by non-aboriginal people whose knowledge is based largely on European philosophies. (Department of Culture and Communications, Government of the Northwest Territories, 1991)

In her work describing Maori epistemologies, Anne Salmond (1985) writes, ". . . .the process of opening Western knowledge to traditional rationalities has hardly yet begun" (p. 260). Here is the problem: People are now seeking to integrate indigenous knowledge and science in the Arctic, but no one quite knows how.

This research on Inuit knowledge and Arctic science is an attempt to interpret them in ways mutually comprehensible to Inuit and scientists. Scientists are defensive in their stance toward issues such as scientific literacy. Some Arctic scientists are defensive about commentary on how science is conducted and applied in the Arctic. Faced with political statements about integrating indigenous knowledge with science, at least one scientist asked the National Science Foundation's Division of Polar Programs: "What *is* indigenous knowledge?" I have approached the problem of integrating indigenous knowledge and science through philosophy of science precisely because I want scientists to pay attention to, and grasp, indigenous knowledge in terms they can understand and, one hopes, respect.

I speak herein about the Inuit of the central and eastern Canadian Arctic; other names are used for closely related peoples (e.g., Eskimo, Yup'ik) in Alaska. The governments of the Northwest Territories and Canada administer most of the study area. Its southern reaches are administered by the province of Quebec. The research problem crosses modern boundaries. I hope that this research may be applied to Inuit life in the contemporary Arctic and to the conduct of science there.

Theoretical background

Arguments about rationality and relativism are the philosophical context for comparing Inuit knowledge with Western science. Inuit knowledge is consensual, replicable, generalizable, incorporating, and to some extent experimental and predictive (Denny 1986; Bielawski 1990). Conversely, I have seen no evidence that Inuit used controlled conditions for experiments, nor did they, over time, increase accuracy in measurement. Their knowledge did not comprehensively address universal phenomena beyond their cultural boundaries, nor did they strive for explanation for explanation's sake.

From the standpoint of philosophical realism, both science and Inuit knowledge contribute to understanding the Arctic. Such an approach requires that one accept the natural world as real and amenable to explanation. It holds that the objects of nature exist in and of themselves, were here before science, and will remain regardless of the kind of inquiry directed toward them. This position takes science seriously, as a special form of knowledge different from the indigenous knowledge of societies without science, *and* it allows that such indigenous knowledge can also contribute to understanding the world. Ian Jarvie (1986) argues strongly that anthropology is ideally suited for realist inquiry about science and other forms of knowledge. A realist approach contrasts with the relativist methods more commonly invoked in describing and validating the indigenous knowledge of oral societies. It is a realist position I take in examining the difference between Western culture, which possesses science, and Inuit culture, which does not (Gellner 1985; Jarvie 1986). I ask, What happens to knowledge when one culture rapidly imports the products of science? On this question, philosophy of science remains silent.

Case study: How Inuit construct knowledge

In the winter of 1991, I visited the community of Inukjuak, in Arctic Quebec. The Inuit who live there call the area Nunavik. There I began fieldwork on one of two case studies for this research. Prior to the fieldwork, I reviewed Nunavik oral histories for data pertaining to Inuit knowledge. I am trying to derive Inuit epistemology (the theory or science of the origin, nature, methods, and limits of knowledge) as others have derived epistemology for several non-Western "schemes of human nature and reality" (Overing 1985, p. 17; Borofsky 1987).

The Inukjuak case study is intended to yield data on how Inuit construct knowledge. The setting is the result of an unusual historical event that deposited people in an environment both uninhabited and unfamiliar to them. In 1953 and 1955, the Government of Canada resettled Inuit from the area around Inukjuak to two locations in the High Arctic: Resolute Bay and Grise Fiord. After 20 to 30 years in the High Arctic, many of these people returned to Inukjuak. They have requested compensation from Canada for their removal to the High Arctic a generation ago. To explain their reasons, they have given many structured interviews quite unlike the usual long Inuit narrative. My task is to learn from the relocatees how they solved the problems of living in an environment where they had no previous experience: how they constructed the knowledge necessary to live there practicing a subsistence lifestyle.

Nunavik oral histories are nearly silent on matters of epistemology, except when the context of specific knowledge—the story that is supposed to contain it—has been lost. When people say they cannot remember stories, and "just say it" (Tuniq 1985), they are most direct in describing how they know something. When they feel that they have lost the context for their knowledge, they speak about how they know it. One hunter, when asked how he worked out the location and movements of caribou herds, looked at me oddly and said, "Because we are Inuit, we can do that" (Interview by author, Inukjuak, 1991).

What does "Because we are Inuit" imply? The initial interviews support the conclusions I drew from the oral histories: Inuit knowledge resides less in what Inuit say than in how they say it and what they do. Inuit knowledge contrasts with science in that "pure knowledge is never separated from moral or practical knowledge" (Overing 1985, p. 17; see also Sayer 1984). Fortunately, Arctic ethnography records a great deal of practical knowledge. However, as Fienup-Riordan (1990) comments, we have paid much less attention to *why* Eskimos do what they do. The difficulty in answering questions of "why" and "how" is stressed in the relevant literature. Salmond, Overing, and others emphasize that an openness to other rationalities, a critical understanding of assumptions built into Western epistemology, and linguistic comprehension beyond the norm even for anthropology are required (Overing 1985; Salmond 1985). Also needed are Inuit anthropologists.

Case study: An ethnography of Arctic science

The second case study begins with a significant contrast between science and Inuit knowledge. Inuit do not separate people from nature. Arctic scientists do. As one scientist said to me, "It's a little bit different in archeology and anthropology, but in geology or biology, people are overburden" (background interview, June 1990, Resolute Bay, NWT). The bifurcation between culture and nature, and between social and natural sciences (Margolis 1987), in Arctic research denies the unity of people and nature that one finds in Inuit interpretations of the world and strategies for living in it.

Fienup-Riordan (1990) takes this contrast further in her work on Yup'ik Eskimo ideology. She writes that Western science assumes an inherent difference between humans and animals and focuses on an explanation of their relationship based on this separateness. Yup'ik Eskimos stand this assumption on its head, assuming instead that men and animals are analogically related as human and nonhuman persons. The focus of explanation shifts to the creation of differentiation out of an original unity. This basic difference in assumptions has led to such actions as the government's relocation of Inuit to the High Arctic, without grasping any of the social and natural implications. It has also led to many Western strategies for community development, medicine, education, and justice that have isolated Inuit from at least half of their reality: the natural world. It also led me to my second case study in progress: an ethnography of Arctic science.

Firmly grounded in the Western intellectual tradition, Arctic science nevertheless differs in specific ways from science that is geographically, financially, and culturally closer to its southern support bases. Arctic science is shaped by environment (e.g., low species diversity, polar extremes), history (e.g., four countries conducting science and applying technology in an Inuit cultural setting), politics (e.g., sovereignty in the Arctic), and humanism (e.g., the Arctic in the southern imagination; wilderness conservation). Arctic science attracts some who are drawn to the Arctic's vastness and extremes, and others who find it free of the overburden of a Western-educated public whose taxes pay for the research. Some see it as a playground; others as the key to global research questions. Arctic science is, willingly or unwillingly, being influenced by the social context in which it is conducted (Cruikshank 1981, 1984; Bielawski 1984; Waldram 1986; Johnson 1987; Colorado 1988; Freeman and Carbyn 1988). This influence has thus far

Traditional Resource Management

> Yup'ik Inuit mythology: Ellam Yua (person of the universe) placed soil on top of the rocks, breathed on her and gave her life. Grass, flowers and trees grew. Then he created the four legged, the winged and the finned. The four legged, the winged and the finned created human beings in the image of Ellam Yua to show him respect and honor. Ellam Yua was so honored that he decreed that human beings would forever take care of all living things on Mother Earth.

Yup'ik Inuit, like all other Indigenous Peoples took their sacred responsibility to take care of all living things on Mother Earth seriously. Because of this communal relationship, Native people have developed a management and regulatory system specifically designed to ensure that all of the resources they use are harvested in ways such that the strength of those resources is always enhanced.

The time-tested knowledge and wisdom of years passed demonstrated the best way to manage and regulate the moose, caribou, and other four-legged herds was to allow only the taking of non-breeding young bulls and old cows. Young cows and bulls have natural mortality rates beginning from birth until they reach maturity in strength, size and age. Many of their fatalities, in winter, result from freezing and starvation. The old non-breeding cows no longer benefit the herd because they can no longer produce young.

Today there are other facts facing us that are not being noted. Facts such as: Why are there unusually higher mortality rates of young bulls and cows during the winter time? Why do the present herds now have unusual regular episodes of collapsing when severe icing winters used to be the only major cause of these natural events? Why are there so many wolves?

The answers to these questions can be found in the present federal and state management systems that have been designed to accommodate sport hunting. All of their policies, regulations and decisions primarily target one species: the large, healthy, strong breeding bull. They specify the size of the rack or horns, and even the number of points it must have before one can shoot it. They force people to use this same system when they are hunting solely to put food on their table, not to put a head mount over their fireplace.

So what happens when one kills only the largest, strongest and healthiest breeding bulls of any herd? We are killing the strong genes that need to be

been primarily empirical. Life scientists, for example, recognize Inuit taxonomic data, but only as appended to studies that are classical Western science (Nakashima 1988).

The ethnography of this case study involves documenting customary research behaviors, beliefs, and attitudes of Canadian Arctic scientists. Their research problems, educational background, and initiation rites are all part of the study. Scientists from three research institutions with critical influence in Canadian Arctic science are involved: the Geological Survey of Canada, the Polar Continental Shelf Project, and the Canadian Wildlife Service.

passed on to make our four-legged brothers and sisters healthy and strong. This condition spreads as the less mature and weaker bulls start breeding before they are prime for the task. Their offspring produce even weaker animals. The weaker animals, especially the new babies, become more susceptible to cold, disease and starvation. When animal herds reach a certain level of poor health and strength, the whole herd will collapse and only the strongest will survive.

The number of predators and scavengers are good indicators for determining the health and strength of the herd. When herds grow in a normal, healthy way, the number of predators will increase slowly because more animals do not necessarily mean more food for them. But when the herds increase and are unhealthy, they grow in numbers at a more rapid pace, because the weaker animals are easier to hunt and many die from the elements.

The Yup'ik Elders say that in the Yukon-Kuskokwim Delta, there used to be a herd of sixty thousand caribou that got wiped out in 1920–21 by very severe and long icy winters. The Elders estimated there to be approximately one hundred fifty wolves preying on that herd. That was all that this healthy herd could sustain.

Now we read that the same number of wolves are preying on caribou herds that are substantially smaller in size. In these areas, the sport hunting management and regulatory system is used and this system is the main contributing factor as to why an over population of wolves exists. In other areas, where this management system is being used and the prime hunting resource is moose, the same situation is being created. There is an over abundance of weak moose and an over population of wolves. In these two areas, the wolves are not indicators of the elements of natural causes but the failure of the management and regulatory systems that foreign governments are using.

Killing the wolves is not going to solve the problems but changing the present management system will. The long term benefits of a traditional management system could again ensure, not only healthy and strong moose and caribou herds, but a balanced, strong and healthy ecosystem.

Apangulak Charlie Kairaiuak

(Excerpt reprinted verbatim from *Tundra Times*, August 24, 1994, with permission of the author and editor.)

Summary

Taken together, the case studies should illuminate how Inuit knowledge and Arctic science both contribute to understanding the Arctic. Each is successful in specific realms. Inuit knowledge has both a moral and practical place in the contemporary Arctic (Overing 1985). But no Inuk will abandon her or his antibiotics, electric sewing machine, or snowmobile. An intellectual tradition crossing and integrating Inuit knowledge and Western science does not yet exist, nor is it clear what an Inuit science might consist of in the future (Bielawski 1990). We are not yet able to construe what an Arctic science that integrates indigenous knowledge beyond empiricism would say about the world.

References

Akuliaq. 1967. Letter in the Avataq Cultural Institute Documentation Centre, Montreal.

Bielawski, E. 1984. Anthropological observations on science in the North: The role of the scientist in human development in the Northwest Territories. *Arctic* 37:1–6.

Bielawski, E. 1990, August. Cross-cultural epistemology: Cultural readaptation through the pursuit of knowledge. Paper presented at the Seventh Inuit Studies Conference, Fairbanks, Alaska.

Borofsky, R. 1987. *Making history: Pukapukan and anthropological constructions of knowledge.* Cambridge: Cambridge University Press.

Colorado, P. 1988. Bridging native and western science. *Convergence* 11:49–69.

Cruikshank, J. 1981. Legend and landscape: Convergence of oral and scientific traditions in the Yukon Territory. *Arctic Anthropology* 17:67–93.

Cruikshank, J. 1984. Oral tradition and scientific research: approaches to knowledge in the North. In *Social science of the North: Communicating Northern values*, edited by P. Adams, 22–29. Occasional Publication 9. Ottawa: Association of Canadian Universities for Northern Studies.

de Cotret, R. R. 1991. Letter to the editor. *Arctic Circle* 1:8.

Denny, J. P. 1986. Cultural ecology of mathematics: Ojibway and Inuit hunters. In *Native American mathematics*, edited by M. P. Closs, 129–180. Austin: University of Texas Press.

Department of Culture and Communications, Government of the Northwest Territories. 1991. *Report of the traditional knowledge working group.* Yellowknife, NWT: Author.

Fienup-Riordan, A. 1990, August. A problem of differentiation: Boundaries and passages in Eskimo ideology and action. Paper presented at the Seventh Inuit Studies Conference, Fairbanks, Alaska.

Freeman, M. M. R., and L. N. Carbyn, eds. 1988. *Traditional knowledge and renewable resources management.* Occasional Publication 23. Edmonton: Boreal Institute for Northern Studies.

Gellner, E. 1985. *Relativism and the social sciences.* Cambridge: Cambridge University Press.

Jarvie, I. 1986. *Thinking about society.* Boston Studies in the Philosophy of Science, vol. 93. Boston: D. Reidel Publishing.

Johnson, M. 1987. Inuit folk ornithology in the Povungnituk Region of northern Quebec. M.A. thesis, Department of Anthropology, Institute for Environmental Studies, University of Toronto, Toronto.

Lopez, B. 1986. *Arctic dreams: Imagination and desire in a northern landscape.* New York: Scribner.

Margolis, J. 1987. *Science without unity: Reconciling the human and natural sciences, Vol. 2—The persistence of reality.* Oxford: Basic Blackwell.

Nakashima, D. J. 1988. *The common eider* (Somateria millissima sedentaria) *of eastern Hudson Bay: A survey of nest colonies and Inuit ecological knowledge*. Environmental Studies Revolving Funds Report 102. Lachine, Québec: Research Department, Makivik Corporation.

Overing, J., ed. 1985. *Reason and morality*. London: Tavistock Publications.

Salmond, A. 1985. Maori epistemologies. In *Reason and morality*, edited by J. Overing, 240–263. London: Tavistock Publications.

Sayer, A. 1984. *Method in social science: A realist approach*. London: Hutchinson.

Tuniq, M. 1985. Interview at Wakeham Bay. Avataq Cultural Institute, Oral History Project.

Waldram, J. 1986. Traditional knowledge systems: The recognition of indigenous history and science. *Saskatchewan Indian Federated College Journal* 2:115–124.

16

Understanding northern environments and human populations through cooperative research: A case study in Beringia

Jeanne Schaaf

> Now today, Park-time. How the hell are we going to live? How the hell are we going to survive? (Gideon Kahlook Barr, August 1993)

Gideon Kahlook Barr, Inupiat elder and historian (Figure 16.1) from Shishmaref, Alaska, speaks of his concern for the subsistence lifeway—the survival of his culture—on ancestral land now included within the Bering Land Bridge National Preserve (Figure 16.2). He has carefully preserved details of traditional technologies, beliefs, and values passed on to him from generations of elders, reaching back long before "Stove-time," "Gun-time," and "White Man's Food-time." He has continued to preserve those aspects of his heritage—in particular, oral tradition and subsistence technologies—which he feels are most important for the survival of his people.

> When I began to know Gideon, the culture was changing. The technology was changing. Dog teams disappeared. Gideon kept a dog team in order to both preserve the dogs and the knowledge of harnessing and sled making and hunting and traveling with dogs. And Gideon told me that some day they're going to run out of gasoline and they're going to need these dogs again. He continued to hunt seals with seal nets which is a very complex operation involved in getting the net through one hole in the ice and somehow moving it through another hole in the ice. Some of the holes have to be seal breathing holes. And from there you wait. So not too surprisingly nobody knows how to set a seal net anymore. But Gideon does and he continued to do it to preserve the technique, the technology. I've always seen Gideon as a Native scholar. (David Hopkins 1993)

David Hopkins began his scientific research in the late 1940s on the northern Seward Peninsula of Alaska, where he later met and began his friendship with Gideon Barr. The Seward Peninsula is a remnant of the Bering Land Bridge, a

Figure 16.1. Gideon K. Barr, Sr. at Ublasaun, Bering Land Bridge National Preserve, 1990. Photo by Ted Birkedal.

vast tundra plain linking Siberia and Alaska, exposed when sea levels were lowered during past ice ages. Cultural and biological exchanges continued after the land bridge was last flooded about 9,500 B.P. (Elias et al. 1992). Hopkins's research on the Bering Land Bridge and greater Beringia helped lead to the establishment of Bering Land Bridge National Preserve in 1980 (Hopkins 1967; Hopkins et al. 1982). There has long been interest in the establishment of an international park in the Bering Strait area as well, in recognition of the important shared biogeographic history between the continents.

Northwest Alaska has been the scene of dramatic climatic changes over the past several million years, including the emergence and submergence of a land bridge that extended over 1000 kilometers at its maximum. Arctic ecosystems of today are largely a product of biotic exchanges across the bridge. Migrations of bird species across the Bering Strait probably represent patterns set in land bridge times. Unique events and processes have preserved—buried beneath ancient tephras (volcanic ash layers) and contained in frozen sediments—detailed records of past environments. By understanding this record of changing environments, it may be possible to predict biotic responses to future climate change.

The archeological record is relevant to the study of both human ecology and climate change because it is a rich repository of environmental data. It is a record of changes in human settlement and subsistence patterns that may reflect long-term fluctuations in resource availability. Scientists and local indigenous people

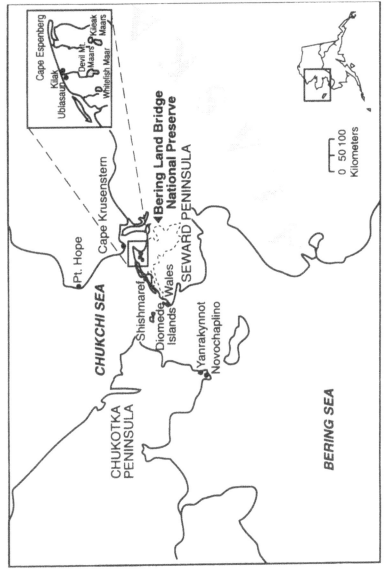

Figure 16.2. Map of Bering Strait area showing locations mentioned in text. Map produced by Theresa Thibault, Alaska Region, National Park Service.

working together on archeological projects have helped bridge the gap between different world views about the value and acquisition of knowledge.

In order to promote scientific exchanges and partnerships, the Alaska Region of the U.S. National Park Service (NPS) initiated the Shared Beringian Heritage Program in 1991 (Schaaf 1992). This program unites Russian and American scientists, land managers, and Native people in a long-term, multidisciplinary research effort focused on the study of traditional lifeways, biogeography, paleoenvironment, and landscape history on the Seward and Chukotka (Russia) Peninsulas. The Bering Land Bridge National Preserve, just 93 kilometers from the Chukotka Peninsula, has been the geographic study area for the initial phases of research. The program is regional and international in scope and includes research within other northwestern Alaska national park lands (Cape Krusenstern National Monument, Kobuk Valley National Park, and Noatak National Preserve) and the Chukotka Peninsula.

Program overview

The Shared Beringian Heritage Program integrates diverse scientific fields on a platform of landscape evolution. Bureaucratic, political, and cultural barriers pose a challenge to the overarching goal of achieving well-integrated studies in this region. Integration of the research is essential because the studies focus on understanding past and present interrelationships of biotic communities, their distribution across the landscape, and changes in environmental relationships over time. The program requires partnerships at multiple levels to achieve these goals.

At the core of the Shared Beringian Heritage Program is a continuation of the partnership that was forged between Gideon Barr and David Hopkins many years ago. Barr is the program's primary Native participant-researcher; Hopkins is the architect of the landscape history research program.

> "One of the things I enjoyed about working with Gideon but also with other Inupiat people is finding out a lot about the landscape that I didn't know to begin with. Of learning nuances that I didn't see." (David Hopkins 1993)

The mutual benefit of close cooperation among scientists, resource managers, and indigenous people associated with protected lands has been widely recognized (Wenzel 1991; McNeely 1992; Chapters 11 and 19, this volume). Many Natives have scientific interests in a variety of disciplines. They possess a detailed knowledge of their environment, based on an intimate relationship with the land and its resources, which has been maintained through many generations (Figure 16.3). Many Native people possess hard data in the form of ledger books, daily diaries, climatic records, photographs, genealogical records, and various artifacts.

Figure 16.3. Two Inupiat children with spotted seal (in autumn near Shishmaref, Alaska, ca. 1923). Photo by Edward L. Keithahn.

This wealth of information remains largely untapped for a number of reasons, including the failure to recognize the depth and validity of Native knowledge (Chapter 15). Sharing personal cultural knowledge with outsiders is sometimes seen by Natives as a loss of intellectual and personal property. Some feel that shared information is often used for the monetary and professional gain of the scientist with little return to the community (Hopkins et al. 1990). Scientific activity tends to be regarded suspiciously in Native communities because it may lead to more agency regulations and restrictions on subsistence activities.

> We have to hunt to survive. They have no sense of seeing that at all. (Gideon Barr, August 1993)

The Alaska National Interest Lands and Conservation Act of 1980 (ANILCA) placed 60% of the land in northwestern Alaska under federal ownership. ANILCA provides a subsistence hunting and fishing priority in the Bering Land Bridge National Preserve and certain other protected areas. Many resource managers and the general public perceive these lands as pristine wilderness, "where man himself is a visitor who does not remain" (U.S. Wilderness Act 1964). Although this perception is changing, there is a failure to realize that these lands are *homelands* that have supported human subsistence activities for 10,000 years. Thus there is a great deal of antagonism between Natives, non-Natives, and resource managers regarding federal management of these Native homelands.

Several projects under the Beringian Heritage Program have made significant progress toward developing real partnerships among park-associated Natives, researchers, and managers. Despite Gideon Barr's concerns about federal interference with subsistence activities, he has worked closely with National Park Service personnel to preserve and record the history of his people. Barr has helped interpret historic and protohistoric sites by sharing details such as names, dates, activities, and oral history (Schaaf 1988). He paints the landscape with a rich layer of place names and associated stories about past events. This is a very visible and viable oral historical record for local Inupiat wherever they look upon the land.

Partnerships in ethnohistory and archeology

> Those old sites [are] so valuable to my understanding 'cause there's something that the ancestors made which we don't even see and use ourselves anymore. That's a valuable part of it right there. How they have survived with their handmade tools, hunting equipment like that in order to survive before white people bring their guns in and their iron and stuff like that. (Barr 1987)

Gideon Barr was the catalyst for the One Man's Heritage Project, an integrated study of the human ecology, ethnohistory, ethnoarcheology, and historic architecture of early twentieth century Inupiat reindeer herders in the Bering Land Bridge National Preserve (Schaaf 1992). Siberian reindeer (*Rangifer tarandus platyrhyncus*) were introduced in northwestern Alaska in 1891 as a source of food after a caribou (*R. tarandus*) decline. Although many Natives participated in reindeer herding, they did not abandon their traditional hunter-gatherer lifestyle, as intended by the government. Rather, herding was integrated into the existing subsistence system. The adoption of herding is a good example of a recent economic and social adaptation to an altered resource base.

Investigations of historic sites associated with the Barr family, who acquired reindeer in 1905, address the question of how the reindeer herding economy was integrated with the subsistence economy of the early 1900s. Archeological research has focused on four historic sites: a winter village (Ublasaun), a fall fishing site, a summer camp, and a reindeer corral constructed in 1919 (Simon and Gerlach 1992). Approximately 20,000 artifacts and faunal remains were recovered and carefully mapped to allow for interpretation of site-use patterns. Oral historical data were collected from elders who provided details of reindeer management and herding practices (Gerlach 1994, personal communication). Details of early herd management show that herding had significant impacts on local wildlife populations and the distribution of natural vegetation in the Bering Land Bridge National Preserve.

Gideon Barr's detailed knowledge of construction techniques, village layout, and site activities facilitated the production of reconstructive drawings of the village. A drawing of Ublasaun as it may have looked during occupation symbolizes the integration of research sought under this program (Figure 16.4). Note the landscape context, wildlife representation, reindeer grazing patterns, details of subsistence activities, and the landward perspective. Earlier archeological maps were drafted from the opposite perspective (from a land-based viewpoint), but Barr, a sea mammal hunter first and herder second, recognized the site only when the map was rotated so the site could be viewed as if it were being approached from the sea.

A temporary laboratory was established in Shishmaref to involve local people in the research program. Technical experts taught classes on a variety of topics including archeology, oral history, scientific illustration, architectural drafting, plant collection techniques, and reindeer biology. Archeological collections from excavations at Ublasaun were initially processed and catalogued by a locally trained lab technician. A local culture historian wrote of the village's response to the program:

> Per the City Council Meeting of the last Wednesday, the activities conducted by your . . . workshops were reviewed. They were very impressed and asked that I request for you to continue the Science Lab Program. I personally feel that your program is more than worthwhile and perhaps actually mandatory. Much of what your program covers is being ignored, not documented and gradually getting lost. The youth interest is especially critical as the parents are not passing on this information (Morris Kiyutelluk Nov. 10, 1992).

This shows great progress in agency-resident relations in an area where merchants distribute calendars that say, "Help preserve our culture, fight the Park Service."

Shishmaref was an important central village in the Tapqaqmuit territory and was probably occupied for hundreds of years prior to that (Ray 1975; Koutsky 1981). Digging by relic collectors has destroyed many of the archeological deposits. During the operation of the laboratory, a number of people brought in their private artifact collections from "Old Shishmaref" for identification, preservation, and photo documentation. The owner of a small ivory figurine was offered $1,000 for it by a buyer from New York, who visits the village every fall (Ningeulook, Hjalseth, personal communications); it would fetch 10 times that amount on the national market. The owner and many others refuse to sell artifacts, because they do not want to see the heritage represented by the artifacts to leave the village.

A partnership research program has been established and funded under the Shared Beringian Heritage Program to encourage village-initiated research in social and natural science. These studies are designed, conducted, and completed by village residents. There are currently five partnership studies in progress: (1) an oral history on community hunting, food preparation, and Eskimo games

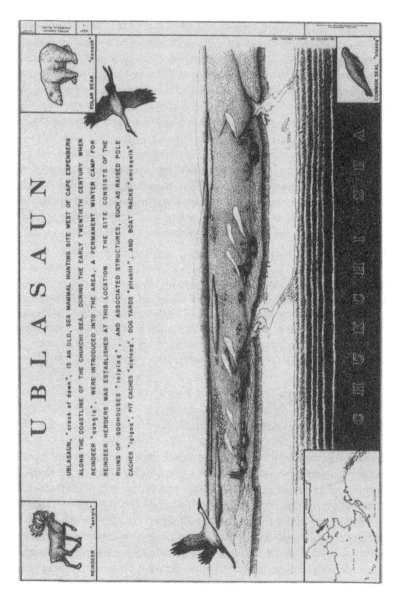

U B L A S A U N

UBLASAUN, "crack of dawn", IS AN OLD, SEA MAMMAL HUNTING SITE WEST OF CAPE ESPENBERG ALONG THE COASTLINE OF THE CHUKCHI SEA. DURING THE EARLY TWENTIETH CENTURY WHEN REINDEER "qungiq", WERE INTRODUCED INTO THE AREA, A PERMANENT WINTER CAMP FOR REINDEER HERDERS WAS ESTABLISHED AT THIS LOCATION. THE SITE CONSISTS OF THE RUINS OF SODHOUSES "inipiaq", AND ASSOCIATED STRUCTURES, SUCH AS RAISED POLE CACHES "igigaq", PIT CACHES "siqiaaq", DOG YARDS "piŋukiit", AND BOAT RACKS "umiaquik".

POLAR BEAR "nanuq"

REINDEER "qungiq"

COMMON SEAL "niqsaq"

C H U K C H I S E A

Figure 16.4. Ublasaun Village, ca.1923. From Historic American Engineering Record, James E. Creech, Alaska Region, National Park Service.

236

(Shishmaref Native Corporation); (2) an oral history on place names and legends (Wales Native Corporation); (3) an oral history on plant, animal, tool, and place names (Brevig Mission Traditional Council); (4) an oral history on winter hunting techniques (Diomede IRA Council); and (5) video documentation of the construction of a traditional *ugruk* (bearded seal [*Erignathus barbatus*]) hunting boat (Kivalina IRA Council). These projects rely on the ethnoscience of community elders and emphasize distribution of the information to community schools and organizations. In addition, the proposed Beringian Heritage International Park and associated scientific information are described for a general audience in the book *Qamani: Up the Coast, In My Mind, In My Heart* (Fair and Ningeulook in press) and in a recently released documentary videotape entitled *Siulipta Paitaat: Our Ancestors' Heritage. Qamani,* coauthored with a Shishmaref resident, presents ethnohistorical research with emphasis on Native contributions.

Another archeological excavation was conducted at a ca. 1800 A.D. village site, separated from the Ublasaun site by 4 kilometers and 120 years (Figure 16.2). The excavation at the Kitluk River Site (Killak) was an opportunity to train local Shishmaref residents in archeological methods in the field and the laboratory. Area elders participated in the identification of the artifacts and interpretation of house architecture.

> Elders are asking what they can get from the past to work with in the future. Archeology like this is recording and preserving the history to teach the children so the culture will not be lost and to upkeep the traditional values. (Fred Tocktoo, crew member, September 1993)

Data recovery from sites of this time period is critical for understanding precontact Inupiaq culture prior to collapse of caribou populations and prior to the advent of reindeer herding, commercial whaling, and widespread availability of European goods. Killak represents a people living in "Bow and Arrow-time" engaged in Siberian trade, on the cusp of rapid cultural change. The investigations at Killak and Ublasaun are providing material data that will directly apply to the investigations of earlier sites close in geographic location, period of occupation, and cultural affinity. The investigations will address the nature of cultural change associated with changing environments, resources, and social conditions in the late prehistoric and early historic periods.

Resource managers and scientists frequently emphasize the Western-based values of knowledge gained through observation, contemplation, and writing rather than activity. Societies with strong oral traditions, such as those of northwestern Alaska, emphasize the development of knowledge as an integrating social activity, with work and communication as major contexts for its expression (Hopkins et al. 1990). Much of the success of the Killak project can be attributed to archeological excavations that helped develop knowledge in ways more closely aligned with the Native world view.

Russian-American collaboration

An anthropological study directed by Russian and American scientists has been recently initiated under the Shared Beringian Heritage Program. Scientists at the University of Alaska (Fairbanks) and Russian Academy of Sciences (St. Petersburg) are documenting early twentieth century linguistic, social, ethnic, and cultural networks between Native people of the northern Seward Peninsula and the northern Chukotka Peninsula (Schweitzer 1993; Schweitzer et al. 1995). Ethnographic research focuses on the infrastructure, demographics, and economy of Naukanski–, Yup'ik–, Chukchi–, and Inupiaq-speaking people of the Bering Strait area. The research documents the history and ongoing intersocietal relations across the Bering Strait within economic, social, and political contexts.

Species lists of birds, mammals, fish, vascular plants, lichens, mosses, and liverworts are being compiled from field data and publications to form the foundation for future studies of central Beringian biota. This effort is being conducted jointly by a team of Native, National Park Service, and National Audubon Society workers. Taxonomic lists are being summarized in English and Russian, with Latin scientific names and English transliteration of Russian names. Siberian Yup'ik and Inupiaq names will be assigned to the lists by Native participants. The Beringian Heritage Program also supports the Russian-American International Pan-Arctic Biota initiative in conducting botanical research in the northern Seward Peninsula, including the development of a botanical database in a standardized format.

Another Russian-American effort focuses on the late Cenozoic geology of the Bering Strait region, with emphasis on glacial impacts on both sides of the Strait. Joint field research is being conducted at the Chukotka Peninsula, Nome, Baldwin Peninsula, Barrow, and St. Lawrence Island. One result of this effort is a new stratigraphic framework for late Cenozoic glacial and sea-level history; many of the geologic sequences are much younger than previously thought (Brigham-Grette and Hopkins 1994). Researchers are investigating the effects of substantial amounts of glacial ice on the land bridge and are rethinking previous interpretations of full-glacial land bridge conditions (Heiser et al. 1992).

As an important outreach activity, the Beringian Heritage Program supported "Crossroads Alaska," a traveling exhibit of artifacts from museums in the Smithsonian Institution, Alaska, and Siberia. The exhibit, as well as educational kits, videotapes, and community programs, are based on themes developed in the joint Russian-United States exhibition "Crossroads of Continents: Cultures of Siberia and Alaska."

Other cooperative Russian-American projects are the development of a Russian-English dictionary of archeological terms and the translation and distribution of Russian documents on Beringia. An international committee of archaeologists is compiling the dictionary, which will contain a Russian-English and

Russian glossary divided into sections on lithic (stone) artifacts, archeological vocabulary, metal and ceramics, and natural science terms in archeology.

Landscape history research

The landscape history component of the Shared Beringian Heritage Program focuses on the dynamic geomorphic environment of Beringia. The first volcanic history research in the Bering Land Bridge National Preserve was conducted by the U.S. Geological Survey in the 1940s (Hopkins 1963). Hopkins' research in Kotzebue Sound and the Seward Peninsula in the 1960s provided the first insights on the paleoecological diversity of Beringia and the potential for tephrochronology (the study of volcanic materials in soils and sediments) to elucidate the late Pleistocene history of the region.

Current research focuses on the development of a comprehensive database on the tephras (ash layers) occurring in the Preserve from 7,000 to 200,000 B.P. (Beget and Mann 1992). Five large maar lakes (lakes formed in rimless volcanic craters) in the Preserve are among the largest on earth (Beget et al. 1995). The Devil Mountain Lake maar (Figure 16.5), the headwaters of the Kitluk River, is 8 km in diameter and is the largest known phreatomagmatic crater (a crater formed by the eruption of steam from permafrost and groundwater, mixed with gases from molten rock). It was formed by an eruption during the last glacial maximum

Figure 16.5. Devil Mountain Lake maar. Photo from Alaska Region, National Park Service files.

about 17,500 B.P., which blanketed the northern Seward Peninsula with a thick deposit of ash (Hopkins 1988).

The paleoenvironment of the land bridge during the last glacial period is being investigated by excavation of the 17,000-year-old ground surface preserved under the Devil Mountain tephra (Hoefle et al. 1994). The preservation of vegetation, faunal remains, and soil across a widespread area affords an opportunity to study the environment of the last glacial maximum in detail. The paleosol (buried ancient soil) characteristics of this region suggest that neither a tundra nor polar desert dominated Beringia following the retreat of glacial ice. Rather, paleoecological data indicate a steppe-tundra (relative treeless plain) that was seasonally water-logged and also seasonally dry during the late Pleistocene on the northern Seward Peninsula.

Mapping of the surficial geology of the Bering Land Bridge National Preserve is also providing a better understanding of the spatial and temporal development of the Beringian landscape. Major geomorphic units are being identified from aerial photographs and from field and laboratory analysis of sediments (Charron and Brigham-Grette 1994). A long sediment core from North Killeak Lake represents a continuous record of paleoenvironmental change over the last 14,000 years (Anderson n.d.). Pollen analysis of the core will define the vegetation history of the area, complement geochemical and sedimentology analyses, and provide a framework for integrating data from other landscape history projects.

The geomorphology of the dynamic coastline of the northern Seward Peninsula is a proxy record of sea-level and climatic history (Mason 1993; Mason and Jordan 1993). Barrier islands (islands built by waves along the mouth of a bay or other indentation along the coast), beach ridge complexes, and sheltered backwater lagoons are important habitat for migratory birds, marine mammals, reindeer, and musk-oxen (*Ovibus moschatus*). Archeological sites on these features spanning 4,000 years also attest to the importance of these habitats for humans. Beach ridges in the region began forming about 4,500 years ago when sea level reached its present position (Figure 16.6). Intervals of increased summer storminess (3,300–1,700 B.P. and 1,200–900 B.P.) are recorded in disconformities of these ridge complexes (Mason and Jordan 1993).

This environmental information helps interpret prehistoric land-use patterns along the coast. Unprotected coastal bluffs provide a source of sediment for deposition elsewhere. Where ground ice is abundant, large blocks break away along the axes of ice wedges, contributing to a high rate of erosion. These eroding bluffs expose ancient coastal plain sediments, valuable in the reconstruction of past Beringian environments. Unfortunately, most archeological sites along the coast are rapidly destroyed by erosion, so that archeological sites older than 400 years are absent along most of the Bering Land Bridge coast.

Native communities such as Shishmaref, whose villages are threatened by catastrophic erosion, are especially interested in the rates of coastal erosion. The

Figure 16.6. Cape Espenberg, Bering Land Bridge National Preserve, 1978. Photo from National Air and Space Administration files.

oral traditions shared by Gideon Barr and other elders preserve a record of storm frequencies, sea-level change, and erosion. According to a widespread folktale, a whalebone monument constructed at the tip of Cape Espenberg by Ilaganiq, an ancestral hunter, records an earlier time when whales were hunted in a bay now too shallow to support more than rare visits from whales (Figure 16.7).

A study of fresh water wetlands or "string bogs" associated with the beach ridge complex at Cape Espenberg is underway. These wetlands are an area of warm permafrost and may be extremely sensitive to temperature and precipitation changes. These wetlands, which contain important wildlife habitat and have sustained human populations over the past 4,000 years, could deteriorate if temperatures increase rapidly. In addition, the wetlands are a reservoir of carbon and methane; increased temperature could enhance the decomposition of organic soils and release greenhouse gases that could accelerate global warming (Chapter 5).

Biological studies

Contemporary studies of the biological environment complement the extensive paleoecological research program in Beringia. For example, a study of vegetation dynamics with respect to reindeer productivity is closely tied to the study of the human ecology of reindeer herding on the Seward Peninsula. Research conducted within the Bering Land Bridge National Preserve is identifying the geographic

Figure 16.7. Ilaganiq's whalebone cairn, Cape Espenberg, Bering Land Bridge National Preserve, 1993. Photo by James Magdanz.

extent of vegetation cover and seasonal ranges for herding. The overall objective of the project is to determine critical habitat requirements of reindeer on the Seward and Chukotka peninsulas and to relate this to animal productivity. Food habits of musk-oxen, another important ungulate species in the preserve, are also being studied. Research is being conducted in cooperation with the Native Reindeer Herders Association of Nome, Alaska, with individual herders as participants.

A final contemporary study focuses on gyrfalcon (*Falco rusticolus*) populations. This large falcon is found in open tundra near rocky outcrops and cliffs on the Seward and Chukotka peninsulas and travels long distances across the Bering Strait. Satellite radio transmitter data indicate that juvenile birds move great distances (3,500 kilometers in 42 days in one case), and that the movements tend toward coastal areas.

Summary

The Shared Beringian Heritage Program has been successful in maintaining close cooperation and integration of research efforts across academic, bureaucratic, political, and cultural boundaries. Adequate long-term funding is a key to supporting field research in remote, roadless areas, and in allowing sufficient time to complete data collection, analyses, and reports. Communication has been critical for integrating diverse elements of the program. Annual meetings for researchers and managers emphasize data sharing. Information is also conveyed through village meetings, lectures, and monthly progress reports. Information reaches a wider audience through professional meetings, publications, and press releases. Participant researchers are keenly aware of the need to demonstrate the value of their research to managers and local residents.

The Shared Beringian Heritage Program has demonstrated that cooperation between nations and cultures can unlock centuries of mystery about the interaction of humans with the Arctic environment. Patterns and processes of the past can provide insight on how climate may affect this interaction in the future. The success of the program may provide a model for future investigations of human-environment interactions in the Arctic and other regions of the world. Through better understanding of the physical, biological, and cultural environment, we may be able to protect resources that were present millennia before the formation of political boundaries.

References

Anderson, P. M. n.d. Pollen Results from North Killeak Lake: A Final Report to the National Park Service. Report on file, NPS Regional Office, Anchorage, 23 pp.

Barr, G. K. 1987, May. Audio taped interview by Jeanne Schaaf. National Park Service, Alaska Regional Office, Anchorage.

Barr, G. K. 1993, August. Videotaped interview by Taylor Productions, Inc. National Park Service, Alaska Regional Office, Anchorage.

Beget, J., and D. Mann. 1992. Caldera formation by unusually large phreatomagmatic eruptions through permafrost in Arctic Alaska. *Eos* 73:6361.

Beget, J. E., S. D. Charron, and D. M. Hopkins. 1995. The largest known maars on earth, Seward Peninsula, northwest Alaska. *Arctic* in review.

Brigham-Grette, J., and D. M. Hopkins. 1994. Last interglacial sea-level record and paleoclimate along the Beringian gateway. *Quaternary Research* in press.

Charron, S. D., and J. Brigham-Grette. 1994. A new glacial to Holocene record from North Killeak Lake, Bering Land Bridge National Preserve, Alaska. *PALEO Times* 2:9.

Elias, S. A., S. K. Short, and R. L. Phillips. 1992. Paleoecology of late glacial peats from the Bering Land Bridge, Chukchi Sea Shelf Region, northwestern Alaska. *Quaternary Research* 38:371–378.

Fair, S. W., and E. Ningeulook. *Qamani: Up the coast, in my mind, and in my heart.* Anchorage: Shared Beringian Program, National Park Service, Alaska Regional Office.

Heiser, P. A., D. M. Hopkins, J. Brigham-Grette, S. Benson, V. F. Ivanov, and A. V. Lozhkin. 1992. Pleistocene glacial geology of St. Lawrence Island, Alaska. *Geological Society of America Abstracts* 24:345.

Hoefle, C., C. L. Ping, D. H. Mann, and M. E. Edwards. 1994. Buried soils on the Seward Peninsula: A window into the paleoenvironment of the Bering Land Bridge. *Current Research in the Pleistocene.* II:134–135.

Hopkins, D. M. 1963. Geology of the Imuruk Lake area, Seward Peninsula, Alaska. *U.S. Geological Survey Bulletin* 1141–C.

Hopkins, D. M., ed. 1967. *The Bering Land Bridge*. Stanford, CA: Stanford University Press.

Hopkin, D. M. 1988. The Espenberg maars: A record of explosive volcanic activity in the Devil Mountain–Cape Espenberg Area, Seward Peninsula, Alaska. In *The Bering Land Bridge National Preserve: An archeological survey*. 2 volumes. Report AR-14. Anchorage: U.S. Department of the Interior, National Park Service.

Hopkins, D. M., and J. G. Kidd. 1988. Thaw lake sediments and sedimentary environments. In *Proceedings of the 5th international conference on permafrost*, 790–795, Trondheim, Norway.

Hopkins, D. M., J. V. Matthews, C. E. Schweger, and S. G. Young. eds. 1982. *The paleoecology of Beringia*. New York: Academic Press.

Hopkins, D. M., W. H. Arundale, and C. W. Slaughter. 1990. Science in northwest Alaska: Research needs and opportunities on federally protected lands. Museum Occ. Paper no. 3. Alaska Quaternary Center, University of Alaska, Fairbanks.

Koutsky, K. 1981. *Early days on Norton Sound and Bering Strait: An overview of historic sites in the BSNC region*. Anthropology and Historic Preservation Cooperative Studies Unit, Occasional Paper 29. Fairbanks: University of Alaska.

Mason, O. K. 1993. The geoarchaeology of beach ridges and cheniers: Studies of coastal evolution using archaeological data. *Journal of Coastal Research* 9:126–146.

Mason, O. K., and J. W. Jordan. 1993. Heightened North Pacific storminess and synchronous late Holocene erosion of northwest Alaska beach ridge complexes. *Quaternary Research* 40:55–59.

McNeely, J. 1992. Nature and culture: Conservation needs them both. *Nature and Resources* 28:37–43.

Ray, D. J. 1975. *The Eskimos of Bering Strait 1650–1898*. Seattle: University of Washington Press.

Schaaf, J. M. 1988. *The Bering Land Bridge National Preserve: An archeological survey*. Research/Resources Management Report AR–14. Anchorage, AK: Alaska Region, National Park Service.

Schaaf, J. M. 1992. The Shared Beringian Heritage Program. *Federal Archeology Report* 5:1–3.

Schweitzer, P. 1993. *Separated neighbors: An anthropological comparison of native communities on both sides of the Bering Strait*. Report on file at NPS, Alaska Regional Office, Anchorage. 56 pp.

Schweitzer, P. P., E. Golovko, and L. P. Kaplan. 1995. Contacts across Bering Strait, 1898–1948. Anchorage: Shared Beringian Heritage Program, National Park Service, Alaska Regional Office. 227 pp.

Simon J., and C. Gerlach. 1992. *Reindeer subsistence, and Alaskan Native land use in the Bering Land Bridge National Preserve, Northern Seward Peninsula, Alaska: A review of the impact of reindeer herding on community relationships and land use in the early twentieth century*. Report on file at the NPS, Alaska Regional Office, Anchorage. 47 pp.

Siulipta Paitaat: Our ancestors' heritage. 1994. Videotape. 29:30 minutes. Shared Beringian Heritage Program, National Park Service, Alaska Regional Office, Anchorage.

Wenzel, G. W. 1991. *Animal rights, human rights: Ecology, economy and ideology in the Canadian Arctic*. London: Belhaven.

17

Biology, politics, and culture in the management of subsistence hunting and fishing: An Alaskan case history

Robert Bosworth

Over the past decade, state and federal laws afforded subsistence harvesting activities a priority over all other consumptive uses of Alaska's fish and wildlife. But a complicated bureaucratic, legal, and regulatory process involving the state and several federal agencies confuses many rural people. Even if legal stability is achieved, questions remain about the ability of resource management institutions to protect a subsistence way of life in Alaska. In a state where commercial fisheries are a $1 billion per year industry, where big game guides charge thousands of dollars per day, and where tourists arrive in unprecedented numbers, subsistence users face increasing competition from higher-profile economic activities.

What is subsistence?

Subsistence use of fish, wildlife, and plants is part of the socioeconomic tradition of many Alaskans (Berger 1985; Schroeder et al. 1987; Caulfield 1991). It is vital to the identity of Alaska Natives and others, and through sharing of foods and the transmission of knowledge, subsistence production provides community cohesion and continuity (Loon 1989; Huntington 1992). In current state and federal law, subsistence is defined as the customary and traditional, noncommercial use of wild resources for a variety of purposes. These uses include harvesting and processing of wild resources for food, clothing, fuel, transportation, construction, arts, crafts, sharing, and customary trade. Food products are generally distributed through noncommercial networks of sharing and exchange. In 1994, the estimated annual wild food harvest in Alaska was 19.9 million kilograms by rural residents and about 4.5 million kilograms by urban residents (Wolfe and Bosworth 1994).

Like many ecological systems, subsistence-based socioeconomic systems develop a stability that enhances their endurance. In both ecological and social systems, diversity is often the key to adaptability and survival in a changing environment. Contemporary subsistence hunting, fishing, and gathering in rural Alaska is part of a mixed subsistence-cash economy in which families use cash income to capitalize in subsistence technologies (Wolfe 1984; Wolfe and Walker 1987). Fishing and hunting are conducted largely by extended family groups who invest in equipment such as nets, boats, snowmobiles, traps, rifles, and chainsaws. Access to innovative hunting and fishing technology is an important adaptive element in subsistence systems. Production is directly consumed and self-limited by the family and community needs. Depending on the region, employment is commonly in commercial fishing, commercial trapping, and the public sector. Mean annual incomes are typically modest and intermittent. The mixed subsistence-cash system also helps to level out the "boom and bust" cycles of the contemporary Alaskan economy.

Fishing and hunting locations are often determined by informal customary rules defining rights of access. Areas such as trap lines and set-net sites are recognized as "use areas" of particular communities and family groups (Brody 1981; Wolfe and Bosworth 1990). Maintaining control over harvest territory is critical to stable subsistence food production. Although more distant seasonal camps may be used for harvesting some species, harvest areas tend to be near communities. Not all of a harvest area may be used each year, or even over a span of years. For example, an area may be hunted only when other areas are inaccessible or when species abundance has changed.

Fish, game, and other foods are typically shared, distributed, or exchanged noncommercially, thus providing much of the social fabric of rural Alaskan communities. Exchanges are often reciprocal and frequently result in labor specialization and efficiency. For example, about 70% of the wild food production in typical rural Alaskan villages is obtained by about 30% of the households (Wolfe 1987; Andrews 1988).

History of subsistence management in Alaska

Although subsistence uses have been recognized in Alaska in federal legislation and international treaties for over a century, legal involvement by the state began in the 1970s following passage of the Alaska Native Claims Settlement Act (ANCSA) in 1971. ANCSA sought to extinguish aboriginal hunting and fishing rights through transfers of cash and land. However, the legislative history of ANCSA reveals that Congress clearly perceived a need for the Department of the Interior and the State of Alaska to further address the issue of subsistence for Alaska Natives.

The first state law pertaining to subsistence (passed in 1974) empowered the state Board of Game to establish subsistence hunting areas, control transportation within hunting areas, and open or close seasons to protect subsistence hunting. The law was never used to establish a subsistence hunting area. The board used other approaches to address subsistence needs, such as winter hunts, which were presumed to benefit local residents. Rural Alaskans, however, increasingly believed that such opportunities would not protect their subsistence needs, and legislation requiring implementation of subsistence regulations by the boards became a political priority.

In 1975, the western Arctic caribou (*Rangifer tarandus*) population crashed. The Alaska Department of Fish and Game (ADF&G) and the board attempted to devise an ad hoc system that would give hunting permits to residents most reliant on these caribou. Three criteria for determining the eligibility of hunters to receive permits were developed: customary and direct dependence, local residency, and availability of alternative resources. (These criteria would reappear several years later in the eventual state and federal subsistence statutes.) This arrangement was challenged in court by a sportsmen's association on the grounds that the board illegally delegated its authority to local communities and that the board did not have the authority to allocate to specific individuals. The Alaska Supreme Court ruled in favor of the plaintiffs; it did not rule on the ability of the board to allocate among individuals.

The Alaska National Interest Lands Conservation Act

During deliberations over ANCSA, Congress recognized that subsistence was an issue of vital importance to Alaska Natives that would need to be addressed in subsequent legislation. Therefore, in the late 1970s, national debate began over subsistence language in the Alaska National Interest Lands Conservation Act (ANILCA, Public Law 96-487). Prompted by the language proposed for ANILCA, the 1978 Alaska subsistence law was a compromise intended to forestall a greater federal presence. For the first time, it directed the Alaska Boards of Fisheries and Game to provide reasonable opportunities for traditional subsistence harvests.

ANILCA was passed in 1980. Title VIII articulated a priority for subsistence use of fish and wildlife and applied specifically to activities on federal lands (60% of Alaska's land base). The authority to manage subsistence uses of fish and wildlife on federal lands was granted to the state, as long as state laws were consistent with federal law. Both state and federal laws defined subsistence uses, stated that subsistence uses would be continued, and created an administrative and legal framework for management, with subsistence as the priority consumptive use.

ANILCA limited subsistence preference to rural residents, but this was not explicitly written into the state law. Following a successful legal challenge to implementation of the state law in *Madison v. Alaska Department of Fish and Game* (696 P.2d 168 [Alaska 1985]), a rural limitation on subsistence eligibility to match the federal law was written into state statute. The revised statute defined a rural area as "a community or area of the state in which the noncommercial, customary, and traditional use of fish or game for personal or family consumption is a principal characteristic of the economy of the community or area."

The McDowell decision

In 1989, subsistence management in Alaska was dramatically changed by the Alaska Supreme Court's decision in *McDowell v. State* (785 P.2d 1 [Alaska 1989]). The court held that "the residency criterion used in the 1986 act which conclusively excludes all urban residents from subsistence hunting and fishing regardless of their individual characteristics is unconstitutional," and that "there are . . . substantial numbers of Alaskans living in areas designated as urban who have legitimate claims as subsistence users. Likewise there are substantial numbers of Alaskans living in areas designated as rural who have no such claims." After the court found that defining subsistence as a "rural" use violated the state constitutional guarantee of equal access to fish and wildlife for Alaska's citizens, the state could no longer implement a subsistence law compatible with federal law because the court's ruling directly conflicted with ANILCA. In 1990, the state was found by the Department of the Interior to be out of compliance with ANILCA, and federal authority was extended to subsistence activities on federal lands.

In the spring of 1990, despite the support of Governor Cowper and the efforts of a Native and rural legislative coalition, the vote on a resolution to place a "rural amendment" to the constitution before the voters failed by a narrow margin. Governor Hickel then convened a citizens' working group on subsistence and brought its recommendations to a legislative session in the summer of 1992. Several measures to resolve the impasse were discussed, including an amendment to ANILCA to make its eligibility criteria conform with the Alaska Constitution. However, regaining full jurisdiction over subsistence management was not a priority of urban legislators, a majority in both the state House and Senate.

"All Alaskans eligible"

A new state subsistence law similiar to the previous one was passed in July 1992. It provided potential subsistence eligibility to any Alaska resident. The priority would apply in areas and for species where the state regulatory boards found that customary and traditional use existed.

The dual subsistence management programs

Dual management began in 1990 when the federal agencies interceded and began to manage subsistence on federal lands. The state was free to maintain its own subsistence program on state and private lands.

The state subsistence program in the early 1990s

By 1990, the state's subsistence management program was making extensive use of the existing hunting and fishing regulatory system, including two regulatory boards. The state Boards of Fisheries and Game had made considerable progress in identifying which uses of fish and game were "customary and traditional" subsistence uses. The heads of local advisory committees in communities throughout the state had been organized into Regional Councils. After a decade of legal challenge and political turbulence, a functioning program had been developed.

The 1992 legislature dramatically altered the meaning of subsistence priority. No longer was the state protecting uses for rural residents most reliant on fish and game. Protection was extended to a much larger group, with the potential for major reallocations of fish stocks and game populations. Consequently, the state regulatory boards became more reluctant to identify the customary and traditional hunts or fisheries because they would be required to provide subsistence regulations, with all Alaskans eligible. With few constraints on subsistence eligibility, the boards perceived that subsistence activities could compete significantly with other uses and would eliminate some hunting and fishing opportunities by nonresidents.

For example, the Nelchina caribou hunt (accessible by road from Anchorage and Fairbanks) was managed prior to 1990 with an allocation of hunting permits to residents of adjacent rural communities and a lottery for remaining permits. Under the 1992 law, any Alaskan could apply for a subsistence hunting permit by answering a questionnaire that was evaluated by three criteria included in the law. (The 1992 law provides a special provision for hunts or fisheries in which subsistence demand exceeds supply.) Those applicants with the highest scores, generally indicating a long family history of dependence on Nelchina caribou for food, would be awarded permits. Because state law restricts participation in any subsistence hunt to residents, the opportunity to hunt caribou from this herd has been unavailable to nonresidents since 1990.

The federal subsistence program

Immediately following the *McDowell* ruling, the directors of federal agencies with land management responsibility in Alaska (U.S. Fish and Wildlife Service, U.S. Forest Service, National Park Service, Bureau of Land Management, Bureau of Indian Affairs) met to consider their responsibilities. The initial federal response to the state's noncompliance was restrained. However, there was a general awareness among federal officials of the need for a major federal subsistence initiative.

Title VIII of ANILCA had been, to the state, a set of guidelines for subsistence management; to federal managers it was a blueprint. The state adapted the preexisting mechanics of fish and wildlife allocation to the requirements of a subsistence priority. The federal government, with no preexisting allocation apparatus, created a framework with subsistence management as its sole purpose. The federal agencies developed a three-part system: (1) a regulatory board devoted to deciding subsistence allocation, (2) 10 regional advisory councils that addressed subsistence issues exclusively, and (3) research and administrative programs that focused on subsistence management.

Parallel subsistence management systems

To rural Alaskans, the most apparent result of dual management was the flurry of regulations to be understood and, if possible, complied with. For example, for most subsistence hunters, a hunting season might begin by finding out whether one's residence had been determined to be rural. Then one would need to know whether the intended hunting area was on state or federal land. Given the patchwork of jurisdictions, hunting might very likely occur on both state and federal land, or even on private land. In any case, a land status map and both a federal and state regulation book would probably be necessary.

In 1994, rural hunters were eligible to participate in subsistence hunts determined by the state Board of Game, or by the federal subsistence board, to be customary and traditional for their communities. For example, a check with the federal regulation book would establish that a hunter in Angoon was eligible to take deer under federal subsistence regulations in Game Management Unit 4 only (identified in regulation as the approximate customary and traditional hunt area for Angoon). The Angoon hunter might choose to hunt outside Unit 4, but would not qualify as a federal subsistence hunter; any hunting there would be under state regulations. State subsistence regulations would probably apply for Unit 4 also, but because all Alaskans were eligible to participate in the state subsistence hunt, the state seasons or bag limits would be more restrictive than federal regulations.

This regulatory maze may be less burdensome to the rural hunter than expected. For most communities, hunting patterns and territories are fairly enduring. If land status boundaries or rural customary and traditional use findings do not change, hunters generally need to learn only once which book of regulations applies to their areas. Hunting requirements are transmitted quickly through social networks, so every hunter does not have to decipher the regulations. State and federal agency staff in rural areas typically help local people understand regulations. If the laws are sufficiently confusing, noncompliance is often the response.

By 1994, rural Alaskans had long felt that their voices too often went unheard in the regulatory board process because of the more numerous urban sportsmen or more politically powerful commercial fishermen. Furthermore, the broad applicability of the 1992 subsistence law, with all Alaskans eligible, convinced many of the futility of depending on the state. Rural Alaskans generally looked favorably on the federal subsistence management system, which addressed subsistence needs exclusively. In summary, the sentiment in much of rural Alaska is that the level of protection to subsistence users provided by the federal subsistence program is greater than the state will provide. Rural people appear to be willing to tolerate the inconveniences of dual management in exchange for this protection.

The resource management institutions: Programs and policies

State wildlife management

The contrast between state and federal management of subsistence is derived not only from legal differences, but also from the history, traditions, personalities, and public perceptions surrounding the state's management of wildlife compared to those of the federal agencies. Hunting and fishing are important parts of the Alaskan way of life. This interest is manifested in a history of citizen involvement in hunting and fishing regulation. State fish and game advisory committees (established in 55 communities) and regulatory boards date back to the early 1960s. The state takes pride in allowing citizens to submit proposals for regulatory change and to speak before the Board of Fisheries or the Board of Game.

Nonetheless, advisory committees have on occasion been dominated by special interests. Several wildlife and fisheries-oriented groups, primarily representing urban sportsmen or commercial interests, have gained political prominence and have developed close relationships with the two boards and ADF&G. It is not unusual for staff from ADF&G, after retirement, to play a significant role in interest groups, community advisory committees, or regulatory boards.

This interplay of political interests, government, and management is the culture of wildlife management in Alaska. It fosters a process of direct citizen involvement in resource decision making and generally achieves a balance of competing resource uses. A conservation ethic has also prevailed, partly due to equilibrium among competing uses and partly due to the principle of sustained-yield management written into the state constitution. But this culture has been stressed by the laws mandating a subsistence priority over other uses rather than balancing competing users.

Federal land and resource management

From Alaska statehood (1959) until 1990, the federal government had limited wildlife management responsibilities. Where substantive federal programs existed, as for marine mammals or waterfowl, they addressed wide-ranging species with national and international management implications. In contrast, the state had a direct hand in the wildlife and fisheries issues affecting most Alaskans.

Federal agencies, however, have played a major role in land management. This has been especially true since the passage of ANILCA. ANILCA's imposition of the national interest in 1980 was a shock to many Alaskans especially to those living in or near the new national parks. From the first discussions of new parklands, it was clear that many Alaskans' traditions of living on the land were antithetical to the National Park Service (NPS) mandate to preserve and protect natural resources, a philosophy cultivated primarily through experience in park management in the lower 48 states.

Suspicion and hostility greeted new Alaska park managers in the early 1980s. In some parts of the state, the basis for this reaction was the experience of rural people with pre-ANILCA Alaska national parks. The relations between pre-ANILCA Glacier Bay National Park and the Huna Tlingit epitomize this type of conflict. The Glacier Bay experience illustrates the systematic process by which NPS has imposed a perceived national interest over the interests of people living in or near national parks.

Management of subsistence use in Glacier Bay National Park

Glacier Bay National Park encompasses an area that is part of the traditional territory of the Huna Tlingit. Evidence of human occupation of Glacier Bay dates to ca. 9,000 B.P., and archeological evidence and oral history indicate continuous use of Glacier Bay by the Hunas and others for at least the past 2,000 years. Such uses include subsistence hunting and fishing, which continued into the 1960s and early 1970s (Ackerman 1968; Schroeder and Kookesh 1988). Subsistence activities in Glacier Bay were actively discouraged by NPS in the 1940s, but it was not until the late 1970s that traditional hunting and fishing were stopped. By that time, language in ANILCA was being developed that allowed subsistence uses to continue in the new Alaska national parks and monuments, but not in the existing parks at Glacier Bay, Katmai, or Denali. Subsistence was a contentious issue, however, because of continued assertion of subsistence rights in Glacier Bay by the Huna Tlingit, reinforced by the State of Alaska's recognition of the Huna's Glacier Bay subsistence fisheries (Bosworth 1988).

In 1991, NPS added to the controversy with draft regulations proposing to eliminate subsistence and commercial fishing in park waters and to allow sport fishing. In 1992, two national environmental groups sued NPS to compel closure of commercial and subsistence fisheries, while the State of Alaska pressed its claim that the waters considered by NPS to be part of Glacier Bay National Park were state-owned. In 1992 and 1993, several bills were before Congress, suggesting amendments to ANILCA to retain traditional fisheries in the park. These efforts, which had been initiated by Alaska's congressional delegation on behalf of Alaska Natives and commercial fishermen, were opposed by several national environmental organizations. In 1994, NPS again initiated discussions about closing Glacier Bay to commercial and subsistence fishing.

Management of subsistence uses in ANILCA national parks

Political instability around Alaska national park management, as described for Glacier Bay, may be what Congress hoped to avoid when it passed ANILCA. Ten new national parks, monuments, and preserves were established in Alaska, with customary and traditional subsistence uses allowed in all of them. The law provided "the opportunity for rural residents engaged in a subsistence way of life to continue to do so" [ANILCA, Sec. 802(1)].

Through ANILCA, NPS found that business would be conducted differently in Alaska. ANILCA not only provided for subsistence hunting and fishing in the new national parks and monuments, but also established local Subsistence Resource Commissions to devise a program for subsistence hunting in the new units. From a local perspective, continued access to harvest territory and effective involvement in subsistence management seemed to have been assured.

But with the passage of ANILCA, some intrusions on subsistence hunting resulted simply from the creation of the new parklands. Publicity brought an increase in tourists. Park rangers began to patrol the parks, and interactions between traditional hunters and new managers were frequently intimidating and occasionally hostile. If there were ever any thoughts that ANILCA would leave local subsistence use unaffected, these were dispelled by new park staff, tourists, planning documents, and public meetings that attempted to reconcile access rights and customary and traditional consumptive use with the preservationist traditions of national park management.

In the second decade after ANILCA, NPS faces two fundamental subsistence policy problems: *growth* in subsistence eligibility and *access* to the parks and park resources. These issues are important to even the most progressive managers, because resolving either one could simplify park management and minimize the need for further subsistence regulation. Other policy issues include concerns over the use of new technology (including transportation technology) associated with subsistence, resource change or damage resulting from subsistence activities, and use conflicts between subsistence harvesters and visitors. The challenge for NPS is to resolve these issues so that fundamental park values are protected while allowing subsistence uses to continue. In terms of the viability of subsistence socioeconomic systems, NPS should be especially concerned about availability of diverse harvest territories, community-based harvest and sharing networks, integrated subsistence-cash systems, and efficient methods of harvest.

Threats to flexibility in use of the harvest territory

Although state and federal subsistence laws provide for continuation of customary and traditional uses, loss of access to traditional hunt areas and fisheries has occurred. The most obvious type of displacement can occur when boundaries used for regulating land uses (such as park or refuge borders) do not coincide with traditional harvest territory boundaries. Access to many types of lands (e.g., game sanctuary, pre-ANILCA national park, privately owned land) can be forbidden regardless of prior use. Dislocation can also occur if state or federal regulatory bodies determine that some areas do not meet "customary and traditional use" criteria. This could occur if use was interrupted for a span of years due to regulation, change in abundance of a resource, displacement, or relocation of a village.

In the new Alaska park areas, displacement of subsistence harvesters can result from interaction with park visitors and from changes in land classification related to park management planning. Conventional recreation management does not recognize the subsistence hunter's perception that an entire river drainage may be unavailable for subsistence use if visitors are camped within it.

Over the past 60–70 years, many indigenous Alaskans have relocated from seasonal camps to central communities. Access to distant portions of traditional harvest areas would therefore be possible only with modern technology such as snowmobiles, small aircraft, or powered boats. NPS regulation of transportation technology to control or prohibit modern equipment can substantially restrict use of harvest territory, reducing flexibility and adaptability of the subsistence system. Similarly, NPS management plans that inflexibly define specific transportation corridors do not allow for necessary shifts in subsistence hunting patterns as habitat and animal populations change. [See Huntington (1992) for a discussion of NPS management of all-terrain vehicles in Gates of the Arctic National Park by Inupiat residents of Anaktuvuk Pass.]

Threats to communities and subsistence-based socioeconomic systems

Subsistence eligibility is the most contentious issue associated with subsistence management in Alaska. The *McDowell* case asserted that the state law illegally disqualified urban hunters. Even prior to the *McDowell* decision, many feared that unrestricted subsistence uses might expand to an unmanageable level. Considerable debate addressed the possibility of developing a system by which state residents might qualify on an individual basis.

In opposition, Native organizations and others argue that subsistence systems are fundamentally organized around the community, not the individual. They point to the existence and importance of community harvest and sharing networks, frequently involving family and extended family groups, which provide much of the unifying fabric of rural communities. They have been resolute in opposing methods of individual subsistence qualification, viewing them as a threat to community-based subsistence systems and traditional Native society. Because of the administrative difficulties in individual qualification approaches, they have also lost their appeal. Few managers have wanted to implement systems allowing neighbors to hunt or fish under different regulations, with the activities of some having legal preference. Agreement has never been reached on the criteria appropriate for such a system.

Concern about growth in the population of eligible subsistence participants in some communities has led NPS managers to suggest systems in which persons would be found eligible to use the parklands on an individual basis. Debate continues within NPS over the "subsistence roster" concept. Some NPS managers favor the approach used since 1980 that established eligibility zones, wherein residents of specified communities had subsistence privileges in parks regardless of individual circumstance. The outcome of this debate will have substantial implications concerning viability of subsistence socioeconomic systems in and near ANILCA parks.

The interaction between the mixed subsistence-cash economic system and government institutions provides additional potential for disruption to subsistence. For example, traditional subsistence activities often include substantial trade in food products. This trade may be considered commercial and therefore prohibited or regulated. Limiting options for generating cash through regulation of customary trade or other means may force communities to develop other methods for creating income. This can include private resource development, such as mining or timber harvesting, which may ultimately be incompatible with subsistence. Such economic development activities can also conflict with adjacent federal, state, or private land management objectives.

Summary

Adaptive strategies inherent in subsistence-based socioeconomic systems account for the endurance of those systems. This endurance requires that the institutions managing subsistence activities do not substantially alter attributes responsible for system survival. These attributes include diverse harvest territories, community-based sharing networks, mixed subsistence-cash economies, and the acquisition and use of new technology.

Several federal agencies assumed subsistence management responsibilities in 1990. Following a strict interpretation of ANILCA, the federal subsistence program established its own policies. Among the federal agencies, all but NPS have missions that are compatible with traditional consumptive uses. Continued subsistence hunting and fishing on the new park lands represents a significant liberalization of NPS traditional management policies. This new policy presents NPS with a range of eligibility, access, and consumptive use issues that are largely foreign to its organizational culture.

Alaska National Interest Lands Conservation Act

The origins of the Alaska National Interest Lands Conservation Act (ANILCA) date back to the late 1950s, when the Territory of Alaska became the 49th state of the United States. The Alaska Statehood Act authorized the State of Alaska to select 42 million hectares (as an economic base) from the total land area of 153 million hectares. This began a lengthy process of withdrawals, selections, and redesignations of lands by state, federal, and Native organizations.

In 1971, Congress passed the Alaska Native Claims Settlement Act, which recognized and settled the long-contested rights of Natives by granting them rights to select 18 million hectares of federal land in Alaska. This act also authorized the Secretary of the Interior to designate new natural, cultural, recreational, and wildlife areas in Alaska. Section 17(d) authorized the Secretary to withdraw 33 million hectares of land to be studied for possible additions to the National Park, National Wildlife Refuge, National Wild and Scenic River, and National Forest Systems. ANILCA, which was passed in 1980, emphasized the following:

- Ten new units in the National Park System (15.7 million hectares). In addition, 2.4 million hectares were added to existing units of the National Park System in Alaska. Hunting is permitted in areas designated as national preserves, and subsistence uses by local residents are continued in national preserves, parks, and monuments designated by the Act.
- Nine new units in the National Wildlife Refuge System (10.8 million hectares; additions to existing refuges of 11.3 million hectares).
- Creation of the Steese National Conservation Area (0.5 million hectares) and the White Mountain National Recreation Area (0.4 million hectares).
- Creation of Misty Fjords National Monument (1.2 million hectares) and Admiralty Island National Monument (0.4 million hectares). Addition of 0.8 million hectares to Chugach National Forest and 0.6 million hectares to Tongass National Forest.

The proposition that government can provide protection to ensure the viability of subsistence-based social systems is now problematic. Several initiatives developed by NPS since the passage of ANILCA have either proposed or implemented restrictive subsistence regulations and policies. These restrictions relate primarily to subsistence eligibility, access, and the use of transportation technologies. The cumulative effect of these restrictions could be a loss of flexibility and adaptability in the subsistence systems associated with the respective parks.

The collision of the interests of subsistence harvesters with those of sportsmen's groups, commercial fishing interests, nonconsumptive recreation users, and government institutions has been a reality in Alaska since long before the

- Designation of segments of 20 rivers within units of the National Park and National Wildlife Refuge Systems as Wild and Scenic Rivers. Seven new segments of rivers were also designated outside National Park and National Wildlife Refuge System units.
- Designation of wilderness for 13.4 million hectares in the National Park System, 7.6 million hectares in the National Wildlife Refuge System, and 2.2 million hectares in the National Forest System.
- Provision for local and regional participation through six regional advisory committees composed of residents of the regions. In national parks and monuments in which subsistence uses are continued, subsistence resource commissions are established to make recommendations to the Secretary of the Interior and the Alaska Governor for a program for subsistence hunting.
- Allowance for customary and traditional subsistence uses by rural residents of wild, renewable resources for direct personal or family consumption on almost all lands (including national parks) affected by the Act. Declared that management of each unit should have the least impact possible on subsistence-dependent rural residents, while conserving fish and wildlife populations. In the event of restrictions, "nonwasteful" subsistence uses have priority over other consumptive uses.
- Authorization of valid existing rights of access and use by snowmobiles, motorboats, airplanes, and nonmotorized surface transportation for traditional activities on conservation unit lands, subject to reasonable regulations. The use of airplanes for subsistence harvest of fish and wildlife is prohibited on all National Park Service lands.

These and other components of ANILCA govern the management of natural resources and subsistence use on most federal lands in Alaska. Although ANILCA was intended to clarify land use issues in Alaska, it continues to be a source of controversy throughout the state.

Darryll R. Johnson

passage of ANILCA. Indeed, recent dilemmas in implementing subsistence management in Alaska are only the most recent manifestation of a conflict that is two centuries old. Current conflicts over subsistence hunting and fishing in Alaska are occurring at a time when demographic change, politics, and technology have introduced enormous complexity into resource management. Competition between conflicting users has never been as intense. Satisfactory stable solutions have never appeared so elusive.

Natural resource agencies are looking for innovative ways to resolve modern biosocial problems. Awareness by managers of the social aspects of resource management coincides with substantial strides by Alaska Natives toward self-determination and an increasing interest in the modern application of traditional ecological knowledge. Greater public involvement in resource decision making, and concepts such as cooperative or comanagement agreements (Chapters 13 and 14), increasingly enter management discussion of policy response to larger issues, including climate change. There is precedent in Alaska for the successful application of cooperative and comanagement mechanisms; the Alaska Eskimo Whaling Commission, Eskimo Walrus Commission, Yukon Delta Goose Management Plan, and Kilbuck Caribou Management Plan are often cited as recent, successful examples of different types of comanagement agreements in Alaska. These approaches provide the best hope for breaking the current subsistence management gridlock.

References

Ackerman, R. E. 1968. *Archeology of the Glacier Bay region, Southeast Alaska.* Laboratory of Anthropology Report of Investigations 44. Pullman: Washington State University.

Andrews, E. 1988. *The harvest of fish and wildlife for subsistence by residents of Minto, Alaska.* Division of Subsistence Technical Paper 137. Juneau: Alaska Department of Fish and Game.

Berger, T. R. 1985. *Village journey: The report of the Alaska Native Review Commission.* New York: Hill and Wang.

Bosworth, R. G. 1988. Consistency and change in subsistence use of Glacier Bay, Alaska. In *Proceedings of the Second Glacier Bay Science Symposium,* edited by A. M. Milner and J. D. Wood, 101–107. Anchorage: Alaska Region, National Park Service.

Brody, H. 1981. *Maps and dreams.* New York: Pantheon.

Caulfield, R. A. 1991. Alaska's subsistence management regimes. *Polar Record* 28:23–32.

Huntington, H. P. 1992. *Wildlife management and subsistence hunting in Alaska.* Seattle: University of Washington Press.

Loon, H. 1989. Sharing: You are never alone in a village. *Alaska Fish and Game* 21:34–36.

Schroeder, R. F., and M. Kookesh. 1988. *Subsistence harvest and use of fish and wildlife resources by residents of Hoonah, Alaska.* Division of Subsistence Technical Paper 142. Juneau: Alaska Department of Fish and Game.

Schroeder, R. F., D. B. Andersen, R. Bosworth, J. M. Morris, and J. M. Wright. 1987. *Subsistence in Alaska: Arctic, interior, southcentral, southwest, and western regional summaries.* Division of Subsistence Technical Paper 150. Juneau: Alaska Department of Fish and Game.

Wolfe, R. J. 1984. Commercial fishing in the hunting-gathering economy of a Yukon River Yup'ik society. *Etudes Inuit Supplementary Issue* 8:159–183.

Wolfe, R. J. 1987, March 12–13. The Super Household: Specialization in subsistence economies. Unpublished paper presented at the 14th annual meeting of the Alaska Anthropological Association, Anchorage, AK.

Wolfe, R. J., and R. G. Bosworth. 1990, May. Territory, subsistence use area, and ecological range: The jurisdictional vulnerabilities of indigenous hunter-gatherer groups in Alaska. Paper presented at the Fifth International Conference on Hunter-Gatherer Societies, Fairbanks, AK.

Wolfe, R. J., and R. G. Bosworth. 1994. *Subsistence in Alaska: 1994 update.* Juneau: Division of Subsistence, Alaska Department of Fish and Game.

Wolfe, R. J., and R. J. Walker. 1987. Subsistence economies in Alaska: Productivity, geography, and development impacts. *Arctic Anthropology* 24:56–81.

18

Biosphere reserves: A flexible framework for regional cooperation in an era of change

Martin F. Price

The concept of biosphere reserves was first proposed as part of the plan for a Man and the Biosphere (MAB) program, submitted to the General Conference of the United Nations Educational, Scientific, and Cultural Organization (UNESCO) in 1970 (UNESCO 1970). In the time since the MAB program was started in 1971, the concept has evolved—both through practical application and in response to trends in the fields of conservation and resource management. Two important trends have been the evolution of the concept of sustainable development and the recognition that local people should be involved in the management of protected areas.

Biosphere reserves have three characteristics that differentiate them from other protected areas. First, they are part of an international system of protected areas designated by UNESCO, rather than by national governments. Second, their outer boundary is flexible rather than being legally defined. Third, the land they contain is administered and managed by more than one agency or owner. In fact, only the first of these characteristics defines all biosphere reserves. However, the second two are key to the concept, particularly when dealing with an uncertain future.

This chapter explores possible roles for biosphere reserves in an era of climate change, with particular reference to North America, especially its far northern latitudes. It begins with a discussion of the evolution of the concept, focusing on the involvement of multiple entities in decision making in the outer zone of biosphere reserves. Then case studies of existing and potential biosphere reserves in North America are presented, and conclusions are offered with regard to the

Discussions and correspondence with many individuals have been valuable in preparing this chapter. I thank Howard Alden (Colorado State University), Bob Alexander (USGS), Bill Gregg (National Biological Service, Washington, D.C.), Hubert Hinote (SAMAB), Fred Roots (Environment Canada), Roger Soles (U.S. MAB Program), Merv Syroteuk (Waterton Lakes National Park), Karen Wade (Shenandoah National Park), and all the participants at the Pack Forest workshop in October 1993. I also thank Paul Brignall for preparing Figure 18.3.

application of the concept. The focus throughout is on values of biosphere reserves for human societies.

Biosphere reserves: A changing concept

The biosphere reserve concept was first formalized in 1973 as part of the MAB program on the conservation of natural areas (UNESCO 1973). Scientific discussion centered on biological diversity. However, it was recognized that greater emphasis should be given to the human uses of the reserves. A subsequent UNESCO task force developed three primary objectives for biosphere reserves: (1) to conserve the diversity and integrity of biotic communities of plants and animals within natural and seminatural ecosystems, including those maintained under long-established land use, and to safeguard the genetic diversity of species; (2) to provide areas for ecological and environmental research; and (3) to provide facilities for education and training.

Two general models for biosphere reserve zonation were developed (Figures 18.1 & 18.2). The first of these included a strictly protected core zone and two buffer zones (administrative units subject to change as needs arise). The inner buffer, with a delineated boundary, was for research and education, with limited public access. The outer one, with a flexible outer boundary, could be used for various purposes, including public recreation controlled according to the carrying capacity of the area (UNESCO 1974). The second model recognized that the optimal approach might be to establish multiple core areas surrounded by buffer zones.

By 1981, 208 biosphere reserves had been designated in 58 countries. However, the conservation role was dominant, and the logistic and development roles were largely forgotten (Batisse 1986). Most reserves had been superimposed on existing protected or research areas, and the idea of formal buffer zones involving other administrative entities was rarely implemented. For instance, 31 of the 38 reserves in the United States were on National Park Service (NPS) or Forest Service (USFS) land; the rest were experimental or protected areas managed by other agencies and institutions (Franklin 1977, Peine 1985). Some reserves were grouped to form clusters, pairing large national parks or monuments with smaller experimental watersheds in the same biogeographic region. The cluster concept acknowledged that few suitable contiguous core areas and buffer zones for manipulation/experimentation or reclamation/restoration were available, and that it would be difficult to harmonize the widely divergent management policies of federal agencies. In Canada, there were only two biosphere reserves: Waterton Lakes National Park and Mont-Saint-Hilaire (owned by McGill University).

There were many difficulties in integrating biosphere reserves into regional contexts involving diverse administrative agencies, landownerships, and land uses (Miller 1983). Although the idea of outer buffer zones was discussed, it was

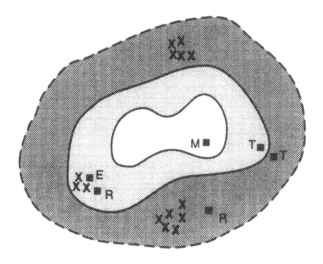

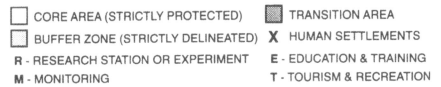

☐ CORE AREA (STRICTLY PROTECTED)   ▓ TRANSITION AREA

☐ BUFFER ZONE (STRICTLY DELINEATED)   **X** HUMAN SETTLEMENTS

R - RESEARCH STATION OR EXPERIMENT   **E** - EDUCATION & TRAINING

M - MONITORING   **T** - TOURISM & RECREATION

Figure 18.1. Schematic zonation of a biosphere reserve, containing a strictly protected core area and two outer zones (UNESCO 1986).

not implemented. Although one criterion for the choice of the smaller experimental reserves was close association with economic interests and decision makers in the region, there was no public participation related to any U.S. biosphere reserve (Worf 1983).

There was some recognition that responsibilities and interests reach out from protected areas to the surrounding community and adjacent lands (Miller 1983). Biosphere reserves could provide (1) a managerial and institutional framework for working with people and natural resources within ecologically defined regions, (2) the ability to break down traditional barriers between institutions and disciplines and to link science and management, and (3) opportunities to study and demonstrate alternative resource management practices. However, such innovative ideas can cause confrontation with established modes of management, and collaboration was slow to develop. Because scientists were often responsible for activities and budgets, research and monitoring were emphasized rather than planning and management.

Similar differences between concept and practice were apparent outside North America, despite recognition of the need to link conservation with human

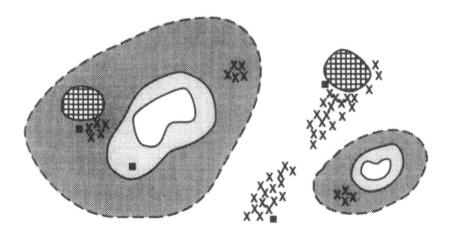

☐ CORE AREA (STRICTLY PROTECTED)   ▓ TRANSITION AREA

▨ BUFFER ZONE (STRICTLY DELINEATED)   ▦ EXPERIMENTAL RESEARCH AREAS

■ RESEARCH OR TRAINING FACILITIES   **X** HUMAN SETTLEMENTS

Figure 18.2. A cluster of biosphere reserves (UNESCO 1986).

activities and rural development (Batisse 1984) and to involve local people in the development of biosphere reserves (Dasmann 1984; Maldague 1984). A Scientific Advisory Panel on Biosphere Reserves recognized the importance of considering local people: "a primary concern of the biosphere reserve is *conservation.* . . . however, . . . the conservation function . . . should be viewed in a more anthropic manner, where *biosphere reserves should be demonstration sites of harmonious, long-lasting relationships between man and the natural environment*" (UNESCO 1986, p. 69, emphasis in original). The objectives of biosphere reserves were redefined into three concerns (UNESCO 1986):

- Conservation—Biosphere reserves should help to strengthen the conservation of biological diversity, genetic resources, and ecosystems.
- Logistics (international research and monitoring)—Together, biosphere reserves should constitute a well-identified international network of areas for research and monitoring, thus providing the needed training and information exchange.
- Development—Biosphere reserves should associate environment and land and water resource development in their research, education, and demonstration activities.

The advisory panel also redefined the outer buffer zone, a transition area or zone of cooperation, as the extent of cooperation between landowners and users of the area and the manager of the protected area (UNESCO 1986). Information deriving from experiments, research, and land management practices within the buffer zone were to be applied in the transition area, thus expanding the sphere of influence of the biosphere reserve. A wide range of cooperative activities were to be developed among researchers, managers, and the local population to promote sustainable resource development while maintaining harmony with the purposes of the biosphere reserve (UNESCO 1986).

Biosphere reserves enable organizations and local people to share expertise, so that resource managers have access to a wider range of knowledge (Gregg 1989; Krugman 1989). However, differences in disciplinary training can lead to contrasting perceptions and conflicts between agencies. Themes of local involvement and interagency cooperation have become an important aspect of protected area design and management (IUCN/UNEP/WWF 1991).

Financial constraints often limit the implementation of biosphere reserves in developing countries where cooperative research, information transfer, and resource management are needed. This is also true in industrialized countries, and the threat of decertification might persuade governments to provide greater resources for adequately implementing the biosphere reserve concept. The threat of climate change may provide impetus for the implementation of biosphere reserves as a component of protected area management.

Regional cooperation: Biosphere reserves in North America

There are six biosphere reserves in Canada and 47 in the United States, including four in Alaska (Table 18.1, Figure 18.3). All but 12 of the 53 reserves are administered by the U.S. National Park Service. Cooperation between United States reserves, even those comprising clusters in a biogeographic province, has been rare. Most do not have all three zones—core, buffer, and transition—and there have been few programs that involved other agencies or local populations.

Waterton Lakes

The Waterton Lakes biosphere reserve (southern Alberta) was designated in 1979, a year after Glacier to the south. Both reserves have core areas protected as national parks. A number of cooperative programs have evolved since 1932 when the two parks were designated as an International Peace Park. However, regional cooperation linked to biosphere reserve designation have been different; while the program at Waterton Lakes has developed substantially, Glacier has had significant problems with interagency relations (Sax and Keiter 1987).

Meetings for local landowners and scientists led to local recognition that the biosphere reserve at Waterton had potential for resolving regional resource man-

Table 18.1. *Biosphere reserves of North America, organized by biogeographic zone*

	Area (ha)	Altitudes (m)
Aleutian Islands		
Aleutian Islands, Alaska[a]	1,100,943	0–3,041
Alaska Tundra		
Noatak, Alaska	3,035,200	30–1,679
Yukon Taiga		
Denali, Alaska[a]	2,441,295	122–6,194
Sitkan		
Glacier Bay–Admiralty Island, Alaska	1,714,851	0–4,666
Oregonian		
Olympic, Washington[a]	363,379	0–2,428
Cascade Head, Oregon[a]	7,051	0–525
California Coast Ranges, California[a]	62,098	0–1,572
Californian		
Central California Coast, California	857,103	−2,300–730
San Dimas, California[a]	6,947	381–1,615
San Joaquin, California[a]	1,832	213–518
Channel Islands, California	479,625	0–741
Rocky Mountains (north)		
Waterton Lakes, Alberta[a]	52,597	1,274–2,918
Glacier, Montana[a]	410,202	972–3,185
Coram, Montana[a]	3,010	1,067–1,920
Yellowstone, Wyoming[a]	898,349	1,710–3,463
Sierra-Cascades		
H.J. Andrews, Oregon[a]	6,100	435–1,631
Three Sisters, Oregon[a]	80,900	600–3,407
Stanislaus-Tuolumne, California[a]	607	1,585–1,951
Sequoia and Kings Canyon, California[a]	343,000	336–4,418
Great Basin		
Desert, Utah[a]	22,513	1,547–2,565
Sonoran		
Mojave and Colorado Deserts, California	1,297,264	−86–3,368
Organ Pipe Cactus, Arizona[a]	133,278	335–1,472
Beaver Creek, Arizona[a]	111,300	900–2,400
Rocky Mountains (south)		
Rocky Mountain, Colorado[a]	106,710	2,328–4,345
Niwot Ridge, Colorado[a]	1,200	2,866–3,780
Fraser, Colorado[a]	9,328	2,660–3,904
Chihuahuan		
Jornada, New Mexico[a]	78,297	1,260–2,830
Big Bend, Texas[a]	283,247	533–2,388
Grasslands		
Central Plains, Colorado[a]	6,210	1,600–1,690
Konza Prairie, Kansas[a]	3,487	320–444
Canadian Taiga/Grasslands		
Riding Mountain, Manitoba	297,591	318–755
Austroriparian		
Big Thicket, Texas[a]	34,217	0–8

	Area (ha)	Altitudes (m)
Central Gulf Coastal Plain, Florida	78,414	0–5
Great Lakes		
Isle Royale, Michigan[a]	215,740	184–425
Long Point, Ontario	27,000	175–188
Eastern Forest (north)		
University of Michigan, Michigan[a]	4,048	183–280
Niagara Escarpment, Ontario	207,240	90–560
Mont-Saint-Hilaire, Quebec[a]	5,550	106–416
Canadian Taiga		
Charlevoix, Quebec	448,000	0–1,170
Lake Forest		
Champlain-Adirondack, New York/Vermont	3,990,000	29–1,629
Hubbard Brook, New Hampshire[a]	3,176	222–1,015
Eastern Forest (south)		
Mammoth Cave, Kentucky[a]	83,377	180–231
Land Between the Lakes, Kentucky[a]	1,560,000	108–201
Southern Appalachian, six states	247,028	226–2,025
New Jersey Pinelands, New Jersey	445,300	0–63
Virginia Coast, Virginia[a]	13,511	0–10
South Atlantic Coastal Plain, South Carolina[a]	8,984	0–100
Carolinian-S. Atlantic, South Carolina	157,105	0–20
Everglades		
Everglades, Florida[a]	592,313	0–2
Greater Antillean		
Guanica, Puerto Rico[a]	4,006	0–228
Luquillo, Puerto Rico[a]	11,340	150–1,080
Lesser Antillean		
Virgin Islands[a]	6,127	0–389
Hawaiian		
Hawaiian Islands, Hawaii[a]	99,545	0–4,170

[a] Administered by one agency

agement issues and that coordination should be a regional activity (Bull 1983). In 1982, a management committee was formed, chaired by local ranchers and including park staff. A technical committee was established with representatives from Canadian and United States agencies administering land adjacent to the park. As the concept was gradually accepted, an unofficial buffer zone/zone of cooperation evolved, including private and government lands within 25 kilometers of the core (national park) area.

From the earliest discussions of the regional value of biosphere reserves, attention focused on organisms that crossed administrative boundaries: beetles (*Dendroctonus* spp.) that attack lodgepole pine (*Pinus contorta*), elk (*Cervus elaphus*) that live mostly in the safety of the park but compete with cattle on adjacent private land, and spotted knapweed (*Centaurea maculosa*) that decreases rangeland quality. Such topics were discussed at public seminars, which sometimes led

Area of Biosphere Reserve

· <20,000 ha

20,000 ha

4,000,000 ha

Figure 18.3. Biosphere reserves in North America.

to research programs with government or private funding (Bull 1983). Consider-able attention was given to public communication, publications, and permanent displays. As local acceptance of the concept increased, the management commit-tee held seminars on more controversial issues (e.g., bear management, game ranching, the sale of provincial land) and organized a demonstration project on brush control. The committee identified management concerns and resolved con-flicts among local ranchers, gas companies, and hunters.

The efficacy of the biosphere reserve increased throughout the 1980s due to involvement of local politicians, industry representatives, park leaseholders, local schools, and park staff (Lieff 1989). Regional biosphere reserve activities were funded through federal and provincial governments and the private sector (Lieff 1989). The scope of these activities has decreased since 1990 because of financial constraints. The management committee still exists, but is involved in only a few projects, and the technical committee no longer functions (M. Syroteuk, 1993, personal communication). The successful application of the biosphere reserve concept is now limited by financial resources, rather than inadequate understand-ing of its potential.

Southern Appalachians

The Southern Appalachian MAB program (SAMAB) was designated in 1988, bringing together two biosphere reserves previously designated in 1976: Great Smoky Mountains National Park (GSMNP) and Coweeta Hydrological Reserve. Together with Oak Ridge National Environmental Park (ORNEP), Mount Mitchell State Park, and Grandfather Mountain (owned by the Nature Conser-vancy), these constitute the core and buffer zones with an area of 247,028 hectares, surrounded by a zone of cooperation with a defined biogeocultural area of 6.2 million hectares (Figure 18.4). Each of the five units is administered by a different agency or institution. SAMAB is formalized in an agreement signed by eight federal agencies as well as the states of Georgia and North Carolina. SAMAB's mission is "to foster harmonious relationships between humans and their environment through programs and projects that integrate the social, physi-cal, and biological sciences to address actual problems" (Hinote 1992, p. 13).

Regional efforts to improve communication between researchers and resource managers already existed in the mid-1970s through annual meetings of four federal agencies and six universities (Franklin 1977). There were also links extending outside federal agencies, such as periodic meetings of conservation organizations to consider resource management issues. In addition, the North Carolina National Park, Parkway, and Forest Development Council was active in linking local NPS and USFS officials with key state and national officials (McCrone 1985).

In 1989, the interagency SAMAB Cooperative was established, and SAMAB was designated as a biosphere reserve. The cooperative established a coordinating

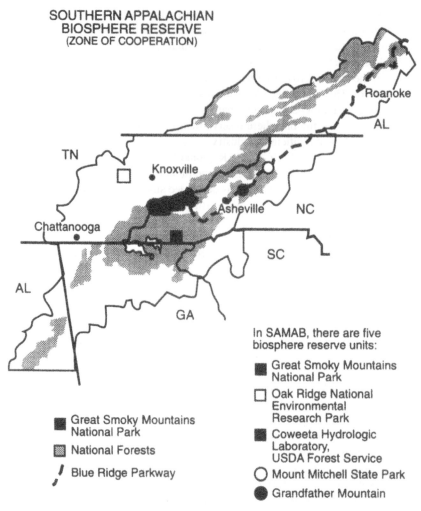

SOUTHERN APPALACHIAN
BIOSPHERE RESERVE
(ZONE OF COOPERATION)

In SAMAB, there are five
biosphere reserve units:

■ Great Smoky Mountains
National Park

□ Oak Ridge National
Environmental
Research Park

■ Coweeta Hydrologic
Laboratory,
USDA Forest Service

○ Mount Mitchell State Park

● Grandfather Mountain

■ Great Smoky Mountains
National Park

▨ National Forests

╱ Blue Ridge Parkway

Figure 18.4. The region of the Southern Appalachians Man and the Biosphere (SAMAB) program.

office, as well as committees on cultural resources, environmental education, environmental research and monitoring, public affairs, resource management, and sustainable development. Members of these committees were drawn from federal agencies, as well as state, local, university, and nongovernmental organizations.

SAMAB has supported a wide range of activities, including research and monitoring, dissemination of information, and the development of policies for air quality and community planning. Many of these activities have required cooperation across administrative/ownership boundaries, such as air pollution, introduction of the red wolf (*Canis niger*), the effects of forest management on neotropical birds, and the spread of dogwood (*Cornus florida*) diseases. SAMAB has been a catalyst for many of these activities, although some might have occurred in the absence of the biosphere reserve because structures for cooperation already existed.

There was initially some mistrust between SAMAB agencies. The agencies differ considerably in their policies and objectives. For example, the USFS perceived that designation, which was spearheaded by NPS actions, tended to emphasize forest preservation over other types of management. Communication within agencies is often less efficient than between individuals with similar backgrounds in different agencies, and funding for joint programs can cause conflicts, because different agencies may not be able to contribute equally.

Activities aimed at solving mutual problems have fostered interagency trust, and SAMAB is increasingly seen as a framework for addressing regional problems. There are now functional networks of people in different agencies. Public trust has been improved by communicating through the media how agencies can cooperate rather than have conflicts. Nevertheless, funding remains a problem; most activities are supported by federal agencies, with few contributions from regional organizations. Community development projects have little support, while traditional activities such as research and education continue to receive funding.

Colorado Rockies

There are three biosphere reserves in the Colorado Rocky Mountains: Fraser Experimental Forest (administered by the USFS), Rocky Mountain National Park (RMNP), and Niwot Ridge (administered by the University of Colorado). Although these reserves (which are not contiguous) were intended to operate as a cluster with the latter two as core areas and the first as an experimental buffer zone, cooperative research and other activities have not occurred.

The idea of a regional biosphere reserve was first initiated by the author in 1989, by inviting representatives of federal agencies (USFS, NPS, U.S. Geological Survey [USGS]), the Colorado Division of Wildlife, and local governments to meet and discuss the situation. The intention was to include the existing three reserves in a regional context with core, buffer, and transition zones. The idea was further developed at meetings attended by representatives of additional agencies and local universities. The idea had considerable support, but progress was slow despite a vision statement explaining that many resources—wildlife, forests, grasslands, air, and water—needed to be considered cooperatively on a regional basis across legal boundaries.

Most agency representatives perceived that a common benefit of a biosphere reserve cooperative would be more efficient exchange of information. The USFS and NPS had medium-term planning needs, and the USGS and Colorado State University were interested in developing regional databases for research and teaching using geographic information systems (GIS). As a result, a data-sharing cooperative was developed in which each participant would provide information about their information database on natural resources.

In 1992, 14 agencies and institutions established a Colorado Rockies Regional Cooperative (CORRC) and an interagency steering committee. The CORRC area of concern comprises parts of five counties, including the three mountain biosphere reserves as well as the Central Plains biosphere reserve (CORRC 1993). The roles of CORRC include: (1) data gathering and sharing, (2) research and management, (3) educational and outreach activities, and (4) affiliation with other programs with compatible or related objectives. Priority issues have been identified through a workshop; a feasibility study; and a survey of citizens, land management agencies, and other organizations.

Unfortunately, CORRC activities are currently restricted to interagency data sharing and research activities. The absence of secure funding has delayed the implementation of an administrative structure and preparation of a proposal to establish a regional biosphere reserve. Only one of the four existing biosphere reserves has joined CORRC.

Regional initiatives: Some conclusions

The examples above illustrate some of the potentials and challenges to establishing a functional biosphere reserve within a regional context. Many other examples could be given [e.g., in developing countries (Brodnig 1993)], and biosphere reserves have also been designated in areas with regional interagency initiatives supported by federal legislation [e.g., Pinelands National Reserve (Good and Good 1984)]. The case studies prepared for the 1993 meeting of U.S. biosphere reserve managers are a particularly good source of information. More systematic surveys of the functioning and regional values of biosphere reserves would be valuable.

A strategic plan for the U.S. biosphere reserve program states that its mission is to develop "a sustainable balance among the conservation of biological diversity, compatible economic use, and cultural values, through public and private partnerships, interdisciplinary research, education, and communication" (U.S. MAB Secretariat 1994, p. 8). This mission should be viewed in the broader context of United States government initiatives toward ecosystem management, a key element of current efforts in resource management on public lands.

Discussions of how to implement ecosystem management continue in different forums (e.g., Keystone Center 1993; Sample 1993), and two areas including biosphere reserves (Southern Appalachians, Yellowstone) have been chosen for demonstration projects. Comparable approaches are being developed outside either the MAB context or government initiatives. These include the Nature Conservancy's "Last Great Places" initiative (Sawhill 1991), programs for bioregions such as the North Cascade Range in British Columbia and the state of Washington (Friedman and Lindholdt 1993), and approaches focused on large watersheds such as the Flathead Valley of Montana (Culbertson 1994).

For biosphere reserves—or other initiatives—to work effectively in a regional context, they should be developed cooperatively by a number of organizations and institutions with adequate funding. The impetus for development should preferably not come from the principal regional land management agency, and many stakeholders should be included from the outset. Stakeholders should comprise not only local agencies and organizations, but also external groups with a strong interest in the region. Unfortunately, most North American biosphere reserves, including those in Alaska, are managed only by federal agencies.

With the exception of Waterton Lakes, the Canadian approach has generally been to take as long as necessary to build regional support for a biosphere reserve before designation is proposed, so that a functional framework (including adequate funding) exists. This approach is also being taken in the Colorado Rocky Mountains and is comparable to what has evolved in Nunavut (Chapter 13). These examples demonstrate the effectiveness of developing a regional approach with a partnership between different systems of resource management—traditional, scientific, and state—with different sets of concepts, terminologies, and decision-making structures.

Time and effort are required to create trust between agencies and organizations. This trust is essential for the effective functioning of a biosphere reserve and is particularly relevant to the inclusion of different stakeholders in a dynamic balance. This has not always been the case in northern Canada (Chapter 13) and has not been attempted in the Alaskan biosphere reserves. Furthermore, local agencies and other institutions must be willing to provide monetary resources because adequate federal, state, provincial, or territorial funding is relatively rare.

Biosphere reserves in the Far North in an era of climate change

There are only four biosphere reserves in northern North America, all in Alaska (Figure 18.3). Biosphere reserves in the far north of Finland, Greenland, Norway, Russia, and Sweden have been the foci for activities of MAB's Northern Sciences Network, which emphasizes the interactions of people and their environments from scientific as well as traditional ecological knowledge perspectives (Chapter 15). The participation of indigenous people in research and management is encouraged (Roots 1993).

This is also the case for the U.S. MAB High-Latitude Ecosystems Directorate project on comanagement of caribou (*Rangifer tarandus*), although this is not within one of Alaska's biosphere reserves. Local people have barely been involved in the management of these reserves, which do not have zones of cooperation. However, some subsistence activities are permitted in the Aleutian Islands and Denali biosphere reserves, and limited cooperation exists between agencies and adjacent landowners. In Glacier Bay–Admiralty Island biosphere reserve, consumptive use of wildlife, which was effectively terminated in the 1960s, is now back on the agenda, as local groups of subsistence hunters and fishermen have developed a coalition with some political strength (Chapter 17).

The minimal local involvement in biosphere reserves of northern North America may change if two current proposals are realized. The first is to create an international biosphere reserve in the Kluane/Wrangell–St. Elias (K/WST) region astride the U.S. and Canadian boundary (Slocombe 1992). This region includes Glacier Bay–Admiralty Island biosphere reserve and the K/WST/Glacier Bay World Heritage Site. Tatshenshini-Alsek Provincial Park (established by British Columbia in 1993) was recently nominated as an addition to the World Heritage Site. Discussions about coordinated management in this region have included officials from NPS, USFS, numerous scientists, and local people. If designated, this would become the largest biosphere reserve in North America. This could be a great value in a region with challenges for developing sustainable resource management of wildlife, fisheries, forestry, tourism, and mining. The jurisdictional issues in this region are complex, but a number of joint management initiatives already exist, and cooperation is developing between agencies at all levels and with local people.

In Arctic Canada, proposals for a biosphere reserve at Igalirtuuq, Baffin Island, have been under consideration since 1989 (F. Roots, 1994, personal communication). In 1991, the community of Clyde River (Chapter 11) approved a conservation plan for a biosphere reserve, which would include both marine and terrestrial components. Its core would be a marine national wildlife area in an area traditionally used for hunting bowhead whales *(Balaena mysticetus)*. Although a draft plan for the reserve was approved, designation in the near future appears unlikely, because Clyde River has no stable source of funding for biosphere reserve activities.

Climate change would add challenges for developing resource management practices flexible enough for changing environmental and social conditions. Regional cooperation among government agencies, nongovernmental organizations, and other groups will be increasingly important. There is no recipe for developing cooperation: The optimal approach will depend on existing relationships between potential stakeholders. In the Far North, stakeholders may increasingly include not only local but regional and national environmental, tourist, and hunting organizations. There are great variations in patterns of land ownership and resource management in the Far North, from the overlapping quilt of jurisdictions in Alaska (Chapters 14 and 17) to the less complex pattern in the Yukon and Northwest Territories.

Issues for which the biosphere reserve concept has been found appropriate are also of concern in a changing climate. Given a suitable framework linking agencies and institutions, regional databases supported by GIS can facilitate development and evaluation of alternative scenarios for resource management. In conjunction with suitable models, GIS can be used to consider many environmental components (location of tree line, fish and wildlife habitat, forest fires, etc.) that may change in response to climate change. These issues could potentially alter the relationships between various economic sectors and subsistence use practices.

Summary

There are challenges to incorporating traditional perspectives, including not just the ecological and economic, but also the social and cultural dimensions of resources (Chapters 9, 11, and 15). Cooperative approaches are essential for resolving potential conflicts with regard to water supplies, habitat for wildlife, use of agricultural land by wildlife, and subsistence harvesting in protected areas. New management strategies may be needed in an era of climate change, especially for migratory populations of wildlife that cross jurisdictional boundaries.

Relatively few North American biosphere reserves, especially in the United States, appear to fulfill MAB objectives relating to sustainable development in the zone of cooperation. Nevertheless, the concept is flexible and powerful, as shown by comparable approaches in the United States and other countries. In an era of climate change, such cooperative approaches also have values for conservation, such as maintaining habitat and corridors for biodiversity and the conservation of migratory and other species (Bridgewater 1991; Gregg 1994). The biosphere reserve model will be increasingly relevant as new strategies are developed in response to changing conditions. The Far North, where diverse comanagement approaches are evolving (Chapter 13)—mostly outside the framework of biosphere reserves—may be one of the best places to realize the potential of the biosphere reserve concept.

References

Batisse, M. 1984. Biosphere reserves throughout the world: Current situation and perspectives. In *Conservation, science and society*, edited by J. A. McNeely and D. Navid, v–xii. Paris: UNESCO.
Batisse, M. 1986. Developing and focusing the biosphere reserve concept. *Nature and Resources* 22:2–11.
Bridgewater, P. 1991. Impacts of climate change on protected area management. *The Environmental Professional* 13:74–78.

Brodnig, G. 1993. The biosphere reserve approach: integration of conservation and development. M.Phil. thesis, Department of Geography, University of Cambridge, Cambridge, UK.

Bull, G. A. 1983. An overview of the situation with respect to Waterton Lakes National Park. In *Towards the biosphere reserve: Exploring relationships between parks and adjacent lands*, edited by R. C. Scace and C. J. Martinka, 22–35. Kalispell, MT: National Park Service.

Colorado Rockies Regional Cooperative (CORRC). 1993. Executive summary of recommendations to establish the Colorado Rockies Regional Cooperative. Fort Collins: Colorado State University.

Culbertson K. 1994, September 28. Balancing tourism development, community values and environmental protection in the Rocky Mountains. Paper presented at Conference on Fragile Environments, People and Tourism, London, UK.

Dasmann, R. 1984. Biosphere reserves and human needs. In *Conservation, science and society*, edited by J. A. McNeely and D. Navid, 509–513. Paris: UNESCO.

Franklin, J. F. 1977. The biosphere reserve program in the United States. *Science* 195:262–267.

Friedman, M., and P. Lindholdt, eds. 1993. *Cascadia Wild: Protecting an international ecosystem*. Bellingham, WA: Greater Ecosystem Alliance.

Good, R. E., and N. F. Good. 1984. The Pinelands National Reserve: An ecosystem approach to management. *BioScience* 34:169–173.

Gregg, W. P. 1989. On wilderness, national parks, and biosphere reserves. In *Proceedings of the symposium on biosphere reserves*, edited by W. P. Gregg, S. L. Krugman, and J. D. Wood, 33–40. Atlanta, GA: National Park Service.

Gregg, W. P. 1994. Potential impacts of climate change and adaptive strategies for natural areas. Washington, DC: U.S. Department of Agriculture. In press.

Hinote, H. 1992, July 13–17. Utilizing biosphere reserves to maintain landscape diversity: The Southern Appalachian experience. Paper presented at Conference on Biodiversity in Managed Landscapes: Theory and Practice, Sacramento, CA.

IUCN/UNEP/WWF (World Conservation Union/United Nations Environment Programme/World Wildlife Fund). 1991. *Caring for the Earth*. Gland, Switzerland: IUCN.

Keystone Center. 1993, November 16–17. Meeting summary, National Ecosystem Management Forum, Airlie, VA.

Krugman, S. L. 1989. Biosphere reserves and the development of sustainable production systems. In *Proceedings of the symposium on biosphere reserves*, edited by W. P. Gregg, S. L. Krugman, and J. D. Wood, 49–52. Atlanta, GA: National Park Service.

Lieff, B. C. 1989. Case study: Waterton biosphere reserve. In *Proceedings of the symposium on biosphere reserves*, edited by W. P. Gregg, S. L. Krugman, and J. D. Wood, 131–141. Atlanta, GA: National Park Service.

Maldague, M. 1984. The biosphere reserve concept: Its implementation and its potential as a tool for integrated development. In *Ecology in practice, part I*, edited by F. di Castri, F. W. G. Baker, and M. Hadley, 376–401. Dublin: Tycooly.

McCrone, J. D. 1985. Biosphere reserves and regional cooperation. In *Proceedings of the conference on the management of biosphere reserves*, edited by J. D. Peine, 46–52. Gatlinburg, TN: National Park Service.

Miller, K. R. 1983. Biosphere reserves in concept and practice. In *Towards the biosphere reserve: Exploring relationships between parks and adjacent lands*, edited by R. C. Scace and C. J. Martinka, 7–21. Kalispell, MT: National Park Service.

Peine, J. D., ed. 1985. *Proceedings of the conference on the management of biosphere reserves*. Gatlinburg, TN: National Park Service.

Roots, F. 1993. Chairman's report on the activities of the MAB Northern Sciences Network. *MAB Northern Sciences Network Newsletter* 13:6–9.

Sample, V. A. 1993. Workshop synthesis: Building partnerships for ecosystem management on forest and range lands in mixed ownership. Washington, DC: Forest Policy Center.

Sawhill, J. C. 1991. Last great places: An alliance for people and the environment. *Nature Conservancy* 41:6–15.

Sax, J. L., and R. B. Keiter. 1987. Glacier National Park and its neighbors: A study of federal interagency relations. *Ecology Law Quarterly* 14:207–263.

Slocombe, S. J. 1992. The Kluane/Wrangell-St. Elias National Parks, Yukon and Alaska: Seeking sustainability through biosphere reserves. *Mountain Research and Development* 12:87–96.

UNESCO. 1970. *Plan for a long term intergovernmental and interdisciplinary program on Man and the Biosphere*. Document 16 C/78, 16th session, General Conference. Paris: UNESCO.

UNESCO. 1973. *Final report, expert panel on Project 8: Conservation of natural areas and of the genetic material they contain*. MAB Report Series 12. Paris: UNESCO.

UNESCO. 1974. *Final report, task force on: Criteria and guidelines for the choice and establishment of biosphere reserves*. MAB Report Series 22. Paris: UNESCO.

UNESCO. 1986. *Final report, ninth session, International Co-ordinating Council of the Program on Man and the Biosphere*. MAB Report Series 60. Paris: UNESCO.

U.S. MAB Secretariat. 1994. *A strategic plan for the U.S. Biosphere Reserve Program*. Washington DC: U.S. MAB Secretariat.

Worf, W. A. 1983. Developing service programs to meet the challenges of biosphere reserves. In *Towards the biosphere reserve: Exploring relationships between parks and adjacent lands*, edited by R. C. Scace and C. J. Martinka, 160–165. Kalispell, MT: National Park Service.

PART V

Essays: Conflict in Social Values

19

Global warming, protected areas, and the right to live off the land

Patrick C. West

As climate change encroaches on northern lands, the interests and traditional life-ways of Native peoples are likely to be threatened. These threats will emanate not just from changes in the natural resource base (e.g., declines in some wildlife populations) but also from policy changes by government agencies such as the U.S. Fish and Wildlife Service, the U.S. National Park Service, and Parks Canada, supported by their constituencies in the conservation community.

Several of the authors in this volume fully understand the attachment of Native peoples to traditional ways of life (Chapters 9, 10, 11, and 16). Norma Kassi (1993) of the Vuntut Gwich'in people (from the community of Old Crow on the Porcupine River in the Yukon Territory) writes of the traditional subsistence-based life of her people, their deep feeling for this way of life, and their dependence on wildlife and other resources:

> Our people are directly affected (by global climate change)....there are no compromises we can make. There are no changes we can make in these old ways. We cannot be compensated for any damage that might occur to our land, the birds, animals, water, fish, and thus our people as a result of another culture's indiscriminations or disregard for others. We have no alternatives to our way of life. This is the only one we know. Without this way of life, we will disappear....I would like to draw your attention to something that is very important to the Gwich'in people....it is called the Porcupine Caribou Herd....on their spring and fall migrations, this large herd walks past our village and onto the Old Crow Flats. They also provide the single, most important food source for the 13 villages where my Gwich'in relatives live. (p. 46)

Increasing numbers of people stand in sympathy with these deep attachments, which represent not only a means of material subsistence but a way of life that deserves preservation—no less than the existence of whales and polar bears (*Ursus maritimus*). But in the face of rapid climate change, reactive and ill-considered natural resource management decisions by the American and Cana-

dian governments are likely. These institutional responses will probably add to
disruption of cultural and adaptive subsistence patterns by Native people unless
corrective structural changes are initiated.

Preservation and the social consequences of cultural disruption

The components of any culture are interrelated, much like the components of an
ecosystem. Cultures evolve to serve certain functions, and those functions give
stability and meaning to the people who participate in them. Many anthropologists emphasize systems of exchange in northern Native cultures in which hunters
share their catch. This sharing is a cultural mechanism helping to ensure group
survival. Groups that failed to develop such a system are not here today, due to
selective survival in the harsh northern environment. In addition, the status that a
hunter attains by sharing is important; without it, feelings of individual purpose
and self-worth may be lost.

Indian and Inuit communities in Canada have repeatedly suffered from an
increase in substance abuse, family breakdown, homicide, and suicide following
government decisions that led to rapid social change and cultural disruption
(Shkilnyk 1985; York 1990). Displacement of the Moose Lake band of the Canadian Cree people when the Grand Rapids Dam was built in northern Manitoba
caused a dramatic increase in social disorganization. Social impacts included a
rise in crime, vandalism, alcohol and drug abuse, collective and individual stress
and anxiety, internal violence, and gang warfare. Prior to construction of the dam
(in the 1960s), the community had no real social problems and a high degree of
coherence (York 1990). The same kind of social disorganization, cultural disinte-
gration, and rapid increase in social problems can occur from displacement when
parks and protected areas are established.

The association between rapid cultural change and the rise of social disorganization is rooted in the breaking of functional relations of culture, social organization, and identity. This breakdown results in widespread *anomie* (the sense of a
lack of identity or purpose in a person or in a society), which leads to alcohol
abuse, spousal and child abuse, family breakdown, suicide, and (in some communities) a higher internal homicide rate. Consequently, an essential element of the
self-determination of Native peoples is the right to decide on the kinds of changes
they will incorporate into their culture, and to adopt these changes at a rate that
will not disrupt functional cultural relationships.

Some environmentally oriented sociologists dismiss such issues by claiming
that the assimilated descendants of today's resource-dependent peoples will look
back on their ancestors' plight as being somewhat quaint (Freudenburg and
Gramling 1994). But these writers depict the cultural transition without the pain,
addiction, and inflicted and self-inflicted deaths that sadly bridge these two eras.

Others dismiss such tragic consequences for indigenous populations by min-imizing their numbers and emphasizing the marginal nature of these cultures to the modern world. At a recent conference on population and the environment, I discussed the horrible consequences some native villagers were enduring—including malnutrition—because of their displacement from a national park in Northern Benin in west-central Africa. One participant commented, "So, what does it matter? These are just a few marginalized people in a modern industrial world." At the time, I was so shocked I could not answer. After some considera-tion, here is my reply: It matters not because it matters to some of us; it matters because it matters to them—to their lives and their hopes for their children, just as our lives and hopes matter to us. These people and their cultures have intrinsic value, as precious as any other component of the environment:

> Many have suffered that we may have the assurance of the preservation of represen-tative global ecosystems. The economic, social, and cultural impacts they have endured have received scant attention and little humanitarian concern on the part of the new ecological mandarins. Our "moral" concern for species everywhere has not extended sufficiently to our own species. Yet they are our common humanity and therefore ourselves....Their children and ours face a common destiny. In our efforts to ensure that they inherit an ecologically viable world let us also strive for an equity that will bind their common humanity in a world where justice, indivisible, shall pre-vail; a world where our children can face their children unashamed. (West and Brechin 1991, p. v)

The previous quote was written in the effort to find a balance between con-servation of protected species and the rights of local native people throughout the world. How much more complicated will these dilemmas become as global warming enters the picture? How much greater will the temptation be for those with power to brush aside polite overtures concerning achieving harmony among conservation, development, and cultural preservation?

Advocacy within the conflict management game

Some advocates for aboriginal peoples shun conflict management as a dangerous strategy in which their interests may be lost. But there are also risks for advocacy groups who do not participate in conflict management. Smart advocates can and should participate by knowing how to play the game and by seeking institutional counterweights to protect their core interests and needs. For parties with less power, such as northern Native peoples, it is especially important that safeguards and counterweights be added to ensure a level playing field (Chapter 12). If con-servationists perceive that their self-interests are at stake in protecting the rights of native peoples, they will be more willing to grant equitable negotiating conditions. They would not be asked to agree with any moral position, but rather to acknowledge the consequences of neglecting the welfare of subsistence-based peoples for their own objectives.

In this volume, Wenzel (Chapter 11) discusses a case in which the rights of Native peoples were overridden. They struck back by sabotaging a protected species. The dominant parties soon realized that because social control and coercion were failing, they would have to involve local people. In the vast regions of the rural North, social control cannot be achieved without high levels of voluntary compliance.

During times of social instability, the price of ignoring the needs of native peoples' needs can be even higher. At the time of colonial transition in India, local peoples ravaged protected area wildlife (Tucker 1991). Recent political unrest in Togo has led to the same result (Lowry and Donahue 1994). Even in times of relative social stability, angry villagers in North Benin have regularly set unwanted forest fires (Hough 1989). Northern indigenous and rural people probably cannot be prevented from harvesting beyond some perceived sustainable level if they choose to do so. If participatory conflict management institutions fail to meet the challenge of ecosystem stress and conflict induced by climate change, actions such as these—clearly counter to conservation interests—are very probable.

The rhetoric of the conservation community is frequently committed to justice for native peoples. But when the going gets tough, will commitment stand? Conservationists have partially failed the test in many areas of the world without the added complication of global warming. They must renew this commitment and convert it to action under the far more trying conditions that lie ahead in the North. Many environmentalists now believe that conservation objectives will be thwarted without the cooperation of local people. Will they revert to older exclusionary behavior if increased temperatures begin to threaten established ecosystems, or will they resist this temptation out of enlightened self-interest?

I carefully laid out the dictum of shared self-interest in a recent presentation about native peoples' rights in parks. At the end of the talk, one bureaucrat said, "I have the answer—guns!" No conflict management regime can exist with this way of thinking.

When push comes to shove: The power equation

Conflict may well increase in the Far North under the influence of global warming, jeopardizing decision-making policies that incorporate the needs of aboriginal peoples (Chapter 10). The greatest danger is not that conflict management institutions will break down, but that parties with greater power (e.g., the federal bureaucracies and environmental interest group constituents) will bend these institutions to their benefit.

When the framers of the United States Constitution sought to build a system of checks and balances, their primary strategy was to decentralize power—neither the Congress, nor the executive branch, nor the courts have sole power. So it should be here. Aboriginal peoples need to create a power base that will resist the tendencies for power to be monopolized by outside interests when protected area ecosystems experience climate-induced stress.

The Havasupai Indians of Grand Canyon National Park provide an instructive example. The Havasupai did not wait for a benevolent National Park Service to build equitable conflict mediation institutions. The Park Service's efforts have been slow and sporadic, and the Havasupai have become proficient in bypassing local park officials in favor of direct contact with the courts and Congress. With land settlement money, they have hired their own lobbyist, gained congressional support, and won several battles (Hirst 1985; Hough 1989). Without the Havasupai's initiative, the present power balance would not exist.

The Havasupai were given property rights under the Grand Canyon National Park Expansion Act, but their land uses (e.g., commercial enterprise) are restricted and are regulated by the Secretary of the Interior. To use their land for any commercial enterprise, the Havasupai might have to fight a coalition of government organizations and their constituencies working together under a system of cooperative domination (West 1994). Cooperative domination exists where shared interests and ideologies between agency bureaucrats and an outside constituency put the bureaucracy in control of the administrative apparatus of the state. Under conditions of climatic warming and both social and ecological instability, it is likely that such a powerful coalition will suppress participatory conflict management mechanisms between parks and resident peoples. Therefore, institution-building between parks and local people is not enough. It is also necessary to codify these procedures in law and to establish a wider regime of checks and balances.

Finding common ground: Regional advisory boards

Another strategy for incorporating checks and balances into the conflict management process is to designate a set of independent regional advisory boards that would stand midway between park/local people relationships and the federal (or other governmental) apparatus. These regional advisory boards would not only bring important indigenous knowledge to the table, but they would assist in providing native peoples with legal counsel and access to technical and scientific expertise to match the legal and technical armory of the protected area administrations and their conservation constituencies. (This was devised by students in the winter 1994 Applied Social and Behavioral Science class in the University of Michigan's School of Natural Resources and Environment.)

Despite the difficulties inherent in predicting climate change and societal response, there is now a growing concern among all stakeholders in the Far North, including Native peoples (Kassi 1993), that a warmer climate might enter the picture and create greater conflicts over subsistence harvesting of wildlife. It is possible that independent regional advisory boards might be able to stimulate facilitation and exchange, through which conservationists, protected area managers, and aboriginal peoples come to see climatic warming as a common threat. If this happens, a united front could be created to deal not only with the impacts of warming, but also with related problems such as carbon output from industrial centers to the south. Aboriginal peoples by themselves would not have much leverage, but they could have greater impact by joining with powerful conservation organizations and park administrations.

Although the best advocacy often occurs in cooperation, the danger of co-optation is very real. Co-optation can be prevented by providing equal access to information and expertise, coupled with genuine recognition and acknowledgment of mutual dependence among the contending parties. For example, all parties may find common interest in the continuity of wildlife populations. All parties also may recognize ultimate moral value in relation to natural resources affected by climate change. Traditional Native hunters, considering themselves part of the sacred animal world, may seek forgiveness from animals' spirits for inflicting their deaths. People of the urban south may seek a secular version of the sacred in attempts to establish and defend protected areas and species. Perhaps global warming will be a threat to that which is sacred to both and will therefore be a path to common ground.

Summary

Aboriginal people in the Far North have a right to live off the land in the face of stress induced by climate change in protected areas. Four things must be accomplished to protect this right. First, there must be increased understanding that the persistence of archaic cultures, labeled as barriers to a more perfect world by outside elitists, is not what is at stake. Ancient *lifeways* are threatened. There must be a recognition of the disastrous social and personal impacts that result from the unraveling of functionally related cultural practices. Out of this must come genuine empathy, not sympathetic nostalgia, for the ways Native people now live. Second, conservationists must act on their rhetoric pertaining to merging natural resource preservation, development, and cultural preservation. Such action will result in more equitable conflict management for Native peoples. Third, conflict management institutions must incorporate systems of checks and balances that will provide a more level playing field for aboriginal peoples. Fourth, institutions such as independent advisory boards can become forums for the recognition of common ground between protected areas and local people. Philosophically, these boards would respect aboriginal rights, while creating a political coalition to collectively combat the impacts of global warming.

References

Freudenburg, W. R., and R. Gramling. 1994. Natural resources and rural poverty: A closer look. *Society and Natural Resources* 7:5–21.

Hirst, S. 1985. *Havasuw 'Baaja: People of the blue green water*. Supai, AZ: The Havasupai Tribe.

Hough, J. 1989. National parks and local people relationships: Case studies from Northern Benin, West Africa, and the Grand Canyon, USA. Ph.D. dissertation, University of Michigan, Ann Arbor, MI.

Kassi, N. 1993. Native perspective on climate change. In *Impacts of climate change on resource management in the North*, edited by G. Wall, 43–49. Department of Geography Occasional Paper 16. Waterloo, Ontario: University of Waterloo.

Lowry, A., and T. Donahue. 1994. Parks, politics, and pluralism: The demise of Togo's national parks. *Society and Natural Resources* 7:321–329.

Shkilnyk, A. M. 1985. *A poison stronger than love: The destruction of an Ojibway community*. New Haven, CT: Yale University Press.

Tucker, R. 1991. Resident peoples and wildlife reserves in India: The prehistory of a strategy. In *Resident peoples and national parks: Social dilemmas and strategies in international conservation*, edited by P. West and S. Brechin, 40–52. Tucson: University of Arizona Press.

West, P. 1994. Natural resources and rural poverty in America: A Weberian perspective on the role of power, domination, and natural resource bureaucracy. *Society and Natural Resources* 7:415–427.

West, P., and S. Brechin, eds. 1991. *Resident peoples and national parks: Social dilemmas and strategies in international conservation*. Tucson: University of Arizona Press.

York, G. 1990. *The dispossessed: Life and death in native Canada*. London: Vintage UK Press.

20

Preserving environmental values in parks and protected areas

Chip Dennerlein

The specter of rapidly increasing temperatures during the next century has been pointed out by scientists throughout the world. In some cases, governments and other institutions have developed plans to mitigate or slow the rate of warming. Across much of the world's far northern regions, climate change may also become a critical matter for indigenous peoples. In the coming century, it may have a profound effect on natural ecosystems and subsistence-based socioeconomic systems of the Far North. Unless scientists, environmentalists, government land managers, and Native subsistence users cooperate, some of the prime environmental values of our northern parks and protected areas may not withstand future biological and human stresses.

Crafting a cooperative approach will be a difficult task because the success or failure in preserving environmental values of northern areas will depend on the willingness of each group to draw from the best of their respective traditions while searching for innovations. Each of us will need to confront our deepest and most sincerely held beliefs and the fundamental purposes of our actions. From agency workshops to village council meetings, organizations and individuals will be asked to separate principles from practices, to preserve enduring principles while altering old practices and embracing new ones. If we fail to meet this challenge, the chapters in this and other recent volumes will do little more than document the decline of some of the most magnificent natural and human systems on the planet.

A paradigm shift in the North

The projected rapid rate of climate change (Chapter 2) will confront us with enormous challenges for which our human experience has not prepared us. Climatological changes that would normally occur over thousands of years may take place over a century or less, leaving us with little time to develop appropriate responses. Unfortunately, government institutions change at a slow pace, and much of the traditional knowledge of indigenous peoples is founded on the under-standing of relatively stable natural patterns (Chapter 15).

Past relations between indigenous peoples and government managers have generally been poor, and bureaucratic dominance and arrogance have been high (West and Brechlin 1991). However, more is required to preserve northern areas than a swing of the pendulum toward more meaningful involvement by local peo-ples, although that swing is needed for any hope of success. Climate change may become an additional factor in a shift of paradigms in the Far North, and each stakeholder will need to reassess his/her own thinking if our values of wildness, wildlife, and traditional human ways of northern protected areas are to be sus-tained.

The biophysical changes and philosophical questions posed by climate change will be enough to give anyone a headache. The manager of a northern refuge, with a budget that barely supports basic waterfowl surveys, may wake one morning to discover that a species of Arctic nesting geese has abandoned tradi-tional resting or breeding areas. The park superintendent who has never received adequate funds for wildlife habitat studies may find caribou (*Rangifer tarandus*), Dall sheep (*Ovis dalli*), wolves (*Canis lupus*), and other "megafauna" declining in population or leaving the park entirely. The subsistence hunter may find a new species competing with or displacing a former species in traditional hunting areas. Understanding the *why* of such changes will be difficult, especially given the trends toward government downsizing and shrinking research budgets, as well as the remoteness of the areas themselves.

For government land managers or Native subsistence users, reconciling the philosophical impact of such changes on their respective cultures will be more traumatic. What is the purpose of a wildlife refuge when the "purpose" has just flown away? What happens to traditional government plans when the special val-ues of protected areas are changing? What is traditional hunting when animals are no longer in their traditional locations? What happens to the right to live off the land when the land itself is changed? These are profound questions, and they will demand of us and our human institutions the hard choice that changes in habitat have always demanded of living organisms: move, adapt, or die.

Traditional approaches may prove insufficient or put us in direct conflict with the values we seek to preserve. For example, wildlife managers may con-sider habitat enhancements, such as those employed to attract waterfowl or encourage the growth of browse for ungulates, as tools to preserve declining pop-ulations in specific areas. In northern areas, however, such attempts will largely be futile. This is not only because of the difficulty of logistics in large, remote northern protected areas, but because we also seek to maintain a sense of wild-ness. Modern society seeks to preserve an essence—a wilderness spirit in the Far North—as much as the physical features and tangible resources. Habitat manipu-lation does not lend itself to such a management goal.

Some traditional responses of indigenous people may also prove inadequate. The coming century will not be the first time in the history of North America that indigenous peoples have faced dramatic environmental change. The breakdown of several highly developed cultures, such as the Anasazi of the American Southwest, were triggered by a period of prolonged drought (Watson 1953). The traditional response of hunter-gatherer societies in the Far North to changes in climate or shifting animal populations has been to disperse or move to new territory (Chapters 9 and 11). The peopling of North America happened largely because Native populations followed migrating animals across the Bering Land Bridge. As the climate changed, people pursued a mix of adaptive strategies to survive. Early hunters depleted some wildlife resources already under environmental stress, learned to harvest different species, and relocated to more favorable habitats. These strategies do not lend themselves to the realities of contemporary subsistence-dependent communities and their juxtaposition in urban society.

Protected area designation and new management challenges

Public policies regarding wilderness preservation, wildlife management, and the rights of indigenous people have emerged with real meaning only over the past few decades (Caulfield 1991). These new attitudes have triggered a series of actions that highlight the challenges we face in preserving wildlands, wildlife, and subsistence lifestyles (Chapters 10, 11, and 17). The major effect of climate change in the Far North will be an additional environmental stress to biological and human systems already struggling to cope with the stresses of changing values, mandates, economics, and technologies.

For both the protected area manager and the subsistence hunter, the Far North of today can seem like unfamiliar territory. The political, social, and economic climate—if not the physical climate—has already experienced profound change. The legal landscape has also been substantially altered over the past few decades. Most northern parks and protected areas (federal, state, provincial) have been established or significantly expanded only in recent years. Many were created under new concepts of land and resource protection, with the notion that modern society faced perhaps its last opportunity to preserve significant portions of intact ecosystems and associated wildlife species. Alaska is an example of this effort. Nearly 70% of the U.S. National Park System land area and two-thirds of the U.S. wilderness preservation system land area are located in Alaska. Linked together, Kluane National Park (Yukon Territory), Tatsenshini-Alsek wilderness (British Columbia), and Wrangell–St. Elias and Glacier Bay National Parks and Preserves (Alaska) form the largest contiguous block of officially protected wildlands on earth. In western Alaska, the Yukon Kuskokwim National Wildlife Refuge encompasses nearly 11 million hectares. Lands such as these represent the ultimate image of wilderness to most people.

Coincident with preservation of large tracts of the Far North was the movement to formally recognize the rights—including land ownership and subsistence—of northern Native peoples. In the United States, the Alaska Native Claims Settlement Act (ANCSA) was passed in 1971, and the Alaska National Interest Lands Conservation Act (ANILCA) was enacted in 1980 (Chapter 17). The societal conscience to preserve the northern wilderness and recognize the rights of indigenous people soon converged to form the context within which we must work today. This new context was unfamiliar to both government land managers and Native subsistence users, and left many unanswered questions about resource use and protection.

Most of our great northern parks and protected areas lie adjacent to or encompass whole communities, primarily (but not always) comprising indigenous peoples who own (as individuals or groups) lands within the protected area boundaries and who have certain rights of access and use within the protected landscape. One of the most significant rights is the pursuit of customary and traditional subsistence activities, including hunting and trapping. In Alaska, this right extends to all national wildlife refuges, national preserves, and national parks, with the exception of the core areas of three national parks (Mount McKinley (Denali), Katmai, and Glacier Bay) established early in the century.

This is a radical change in the philosophical, institutional, and legal framework of the U.S. National Park System. In the 368 National Park System units outside of Alaska, hunting is authorized in only a few of the units. Moreover, the status of some areas of Alaska can be a confusing mix of federal lands and privately owned Native corporation lands. The Yukon Kuskokwim Wildlife Refuge may encompass 11 million hectares, but nearly 3 million hectares will ultimately be conveyed to more than 40 Native village corporations. In addition, the villages belong to a regional corporation that will own most of the subsurface estate beneath the village-owned lands. Six million hectares within parks and refuges will be owned by private Native corporations. The real significance of Native landownership in parks and refuges can easily be lost in the land area figures. Every real estate broker knows the three most important rules of real estate value: "location, location, and location." It is the same with natural systems.

Through thousands of years of trial and error, indigenous people learned the most productive areas for hunting, fishing, and gathering. It was simply a case of knowledge needed for survival. When government policies (e.g., universal schooling) ended nomadic or semi-nomadic lifestyles of northern hunter-gatherers and brought about the establishment of permanent villages, Native people chose traditional seasonal village sites as permanent communities. In 1971, ANCSA mandated that the newly established Native corporations select land in predetermined patterns around existing villages. As a result, Alaska Natives gained title to millions of hectares of productive habitat. Nearly a decade later, when the U.S. Congress established new areas and expanded existing parks and refuges in Alaska, Native corporations had long since made their land selections. Today, most Native-owned lands are located along key coastal or riparian zones or the lowlands of mountain valleys. Management of critical habitat in these lands is essential to the long-term environmental health of many parks and refuges. While northern wildlife requires vast, unspoiled landscapes, their survival often depends on the preservation of specific areas. The value of a million hectares can be compromised by the poor management of a thousand.

Extensive parks and refuges in Alaska would not have been politically possible without Native consent and active support. Native leaders understood that while Natives would own some of the best lands for subsistence, the continued viability of subsistence required the utilization of larger areas. Substantial areas of federal lands needed to be reserved in their natural state and protected from certain forms of development or disposal to state or local governments or to private individuals. Advocates for the preservation of wildlands for modern society and advocates for preservation of traditional subsistence lifeways became allies. Forty-five million hectares were set aside in a single law. Those who wanted continued exploitation of the northern landscape recognized the implications of a coalition of environmentalists, federal officials, and Native subsistence advocates—labeling it an unholy alliance.

While the alliance was not unholy, it was fragile. There were certainly environmentalists, government officials, and Natives who had a vision of the rural subsistence community as a compatible, even essential, element in the new protected areas. However, these different cultures did not generally accept or understand each other's interests. Against this background, both the modern society of government land managers and environmentalists, and the subsistence culture of Alaska Natives continued to change.

Indigenous people of the Far North have experienced dramatic changes over the past several decades. There have been improvements in education, economic opportunities, and health care, but stresses on individuals and communities have resulted in high rates of alcoholism and other destructive behavior. Mixed subsistence-cash economies, increased population, and improved technologies have substantially affected subsistence (Jorgenson 1990).

As more villagers move to regional centers, they employ more modern technologies (e.g., powerboats) to travel long distances to hunt. Along the Yukon River, upriver villages now close Native corporate lands to nonresidents (including Native subsistence hunters) wishing to come from downriver villages or regional centers because of increasing competition for wildlife. This is a difficult issue for villagers; no trespassing regulations seem inconsistent with traditional respect for hunting territories or sharing of subsistence resources. However, economics and technology have changed the rules. As the population increases in rural areas, hunters need to travel farther from permanent villages to find game. In Alaska, subsistence rights are not confined to indigenous people, but are granted (on federally protected lands) to all local rural residents. Virtually all citizens qualify under state law on state-owned lands. Over time, the previously clear connection between traditional lifestyles and subsistence use will become increasingly blurred. Unmanaged, these trends can have significant negative effects on the values of parks and protected areas.

Key characteristics of subsistence systems include "diverse harvest territories, community-based sharing networks, mixed subsistence-cash economies, and the acquisition and use of new technology" (Chapter 17). Bosworth further suggests that government management policies and programs must protect these characteristics and minimally impinge on subsistence adaptability. Protecting these subsistence characteristics and preserving the environmental values of protected areas are not mutually exclusive goals. They can even be mutually supporting goals if the participants adhere to certain standards of character and conduct.

Environmental values in protected areas

In the process of reaching successful agreements, what is really negotiated is *meaning*. Environmental values *can* be preserved while protecting a compatible human subsistence ecology. Values do not exist in isolation; they are not determined by a magic data set. Although environmental values of northern parks and protected areas can be quantified, the values themselves are not simply a product of science. They also reflect the meaning of the parks and protected areas in our society. What does modern society expect of these areas to the Far North that we have set apart? In simplest terms, we expect two things: (1) a wild landscape unmarred by machines and human development, and (2) land in which natural and healthy populations of wildlife can be observed roaming free. Sustaining this condition over time will require discipline. As one step, we need to address the use of technology, especially when combined with widespread harvest territories for subsistence. Control of technology has been fundamental to the protection of hunted wildlife populations since the inception of serious wildlife management. Such control is necessary because certain technologies enable destructive behavior and must be prohibited if they severely damage the environment.

Waterfowl hunting in North America provides an example of an appropriate response to these principles. When it was determined that lead shot in shotgun shells was polluting aquatic habitat and poisoning waterfowl that ingested the lead pellets, lead shot was prohibited in favor of (nontoxic) steel shot. Steel shot addressed concerns about environmental quality and enabled hunting to continue, demonstrating that new technology can be a positive force.

It is not the use of technology—new or old, for subsistence or recreation—that is the issue, but the effect of that technology on environmental values. The use of all-terrain-vehicles (ATVs) illustrates this point. Motorboats may break the wilderness silence, but they leave no trace if they do not damage riverbanks or intrude on nesting areas. Similarly, snowmobiles will not harm the land where there is adequate snow cover. However, ATVs leave long-lived physical scars across fragile northern landscapes. Some areas of northern Russia bear stark testimony to the devastation caused by the widespread use of ATVs.

Should ATVs be considered traditional modes of transportation for subsistence activities in parks and protected areas? This question goes to the heart of the meaning of traditional adoption of new technologies. If maintenance of subsistence culture requires contemporary subsistence hunters to travel long distances over tundra by means of ATVs, key characteristics of subsistence are in conflict with some of the values of northern parks and protected areas for modern society. If subsistence hunters respect the character of the land, this difficult issue will be solved. Likewise, we will be able to successfully confront issues of bag limits, no hunting zones, and prohibited species—even the difficult issue of subsistence eligibility—by protecting subsistence traditions based on abundant wildlife in a wilderness landscape.

Humans have not prospered on Earth as a species by saying "no" to technology. Moreover, it is only recently that modern society has perceived that limits are necessary for the preservation of natural resources. If we are to preserve environmental values and traditional human cultures, we must affirm common meaning in our values. Government managers must do more than *accept* subsistence; they must *believe* in it as an ecological component of the Far North. Their role is not simply as managers, but as protectors of a valuable human ecology, one that can be a significant element in the long-term preservation of parks and protected areas.

Subsistence users must do more than simply *accept* parks and protected areas as a necessary evil. They must come to *embrace* the preservation of unmarred landscapes and plentiful wildlife populations as a reflection of traditional subsistence culture. They must put conservation above convenience (in the case of technology), embrace limits on the taking of wildlife and on eligibility for participation (in the face of expanding populations and changing demographics), and confront restrictions on the use of subsistence resources for trade and barter (in the face of increasingly mixed subsistence-cash economies and widespread wildlife markets).

Subsistence and management challenges

Although subsistence is changing, it is vital to many northern communities (e.g., Brody 1987). Beyond its nutritional importance, subsistence is the defining and unifying element of northern indigenous cultures (Chapters 9, 10, 11, and 17). Humans do not easily change their habits of resource use when those habits are linked to identity as well as economy. Witness the bitter controversies over timber harvests in the Pacific Northwest, where socioeconomic values embodied in logging have collided with environmental values of preserving wildlife and ancient forests. Resistance is high, even when solutions are proposed for altering traditional logging practices and developing new economic opportunities while preserving ancient forests. As the battle wages, both environmental values and a way of life are being lost forever.

Several years ago, declines in certain moose (*Alces alces*) and caribou populations in Alaska prompted state wildlife managers to promote predator control, specifically the killing of wolves. The objective was to depress wolf populations to decrease predation and facilitate an increase in caribou populations, with the ultimate goal of greater caribou harvests. Predator control programs are part of the history of wildlife management in the North. At one time, a bounty was paid on the claws of bald eagles (*Haliaeetus leucocephalus*). However, as society changes and becomes more urban, values regarding wildlife also change. The wolf has become a sacred symbol of wilderness not just to animal rights advocates but to the general public. The wolf control program quickly became a statewide controversy and a national debate. Although the program was aimed at providing additional opportunities for urban hunters, there was a foreshadowing of future confict when state wildlife officials initially justified the program on the basis of providing more caribou for subsistence hunters.

Indeed, some subsistence hunters have called for wolf control to help increase herds of both caribou and moose (but see "box" by Kairaiuak, pp. 224–225). In some areas of Alaska, moose have become a more finite subsistence resource in the face of increased competition from an expanding human population. Subsistence hunters formerly spent considerable time trapping as well as hunting. Contemporary subsistence hunters, often with schedules dictated by mixed subsistence-cash economies, do not have the time (or in some cases the interest) for the time-consuming work of trapping. Thus interest in moose hunting continues to increase, while interest in trapping declines, and the subsistence hunter's effect on the natural system is altered. But government predator control is not the answer. Traditional subsistence is in grave danger if it depends on the use of aircraft and other technology to depress competing wildlife species.

Increasing numbers of urban residents are coming in contact with subsistence activities in the Far North through media accounts and by visiting parks and protected areas. While conventional tourism has grown at a rate of 5% per year, nature-based tourism (ecotourism) has grown at an astounding 30%. Although modern visitors may support the general (perhaps romanticized) notion of the role of human subsistence as part of northern areas, these visitors have high standards for wilderness. They will accept subsistence practices that do not readily fit their preconceived notions, *provided* they can also see that such activities do not erode the environmental values of protected lands.

How do we develop our future management policies? Modern society will continue to support subsistence harvest in northern protected areas, but not on the basis of maximum sustained yield. Visitors will expect to witness wildlife in significant numbers, not glimpse reduced populations at the edge of sustainability.

Society will accept the sharing of wildlife for food and its use in crafts, but it will not accept trafficking in wildlife or the harvest and sale of protected species under strained interpretations of traditional barter systems. Society will agree to diverse harvest territories, but not to unlimited travel across fragile terrain with damaging equipment.

Traditional and customary (and acceptable) activities at one level do not necessarily create a traditional (or acceptable) condition at another. Each action may be traditional, but the result of those actions (by increased numbers of participants over a long period) may produce a dramatically altered environmental condition. There is a difference between one ATV and a convoy, particularly over time. Even when prior rights are recognized, they must be circumscribed based on their effect on natural resources. A dense network of ATV tracks through sensitive wetlands cannot be justified on the basis of traditional use. Decisions may become more difficult in the face of climate change. What will happen when local residents (whose village was chosen to provide access to subsistence resources) and government managers are confronted with wildlife populations relocating to distant valleys because of habitat changes?

Northern people can rise to the challenge. Cooperative efforts have produced measurable results in restoring threatened populations of several species of geese on the Yukon-Kuskokwim Delta. The Alaska Eskimo Whaling Commission has blended wildlife conservation, modern regulation, traditional sharing, and inter-village cooperation to protect migrating whales and sustain a vital subsistence harvest. These and other successful emerging models share common traits. They involve the participation of citizen conservationists, government managers, and subsistence users. They are based on agreements guided by science but founded on shared principles. These models must extend from wildlife management to land use, including Native-owned lands (principally along riparian zones) that contain habitat for wildlife that also use parks and protected areas.

Agencies are recognizing the need for more comprehensive ecosystem management, especially where preserving protected areas depends on a conservation ethic on adjacent nonprotected (public and private) lands. Sustaining a viable subsistence culture, even a mixed subsistence-cash culture, may be one of the best ways for government resource managers to preserve the values of parks and protected areas (Chapter 14). Communities that define a significant portion of their identity through the use of wildlife resources are more likely to protect the natural values of their own lands and cooperate in the protection of public lands.

Summary: A sustainable future

Reexamining our respective values and communicating shared meaning will take time, although we may not have the luxury of time. As Wallace Stegner (1992) wrote, "Values, both those that we approve and those that we don't, have roots as deep as creosote rings, and live as long, and grow as slowly." All of us—environmentalists, government managers, Native subsistence users—view the landscape of the Far North from deeply rooted perceptions. But it may be possible for us to find the deeper values that allow us to agree on principles and adjust our practices to preserve those deeper values. We must develop and strengthen these new models before we are forced to face the increased stresses of climate change. We cannot afford a continual legal and political battle over regulations and enforcement actions. Without shared meaning there will be only conflict, and as with all battles, there will be casualties. In this case, the first casualties will be the wildness and wildlife of our northern parks and protected areas. However, traditional human cultures of the Far North will not long survive a battle against the overwhelming forces of a larger modern society. In the end, the most sacred values of northern parks and northern people will be lost. We cannot allow this to happen. We must work toward a sustainable future for the natural *and* human ecology of parks and protected areas of the Far North.

References

Brody, H. 1987. *Living Arctic: Hunters of the Canadian north.* Vancouver, British Columbia: Douglas and McIntyre.

Caulfield, R. A. 1991. Alaska's subsistence management regimes. *Polar Record* 28:23–32.

Jorgenson, J. G. 1990. *Oil age Eskimos.* Berkeley: University of California Press.

Stegner, W. 1992. *Where the bluebird sings to the Lemonade Springs: Living and writing in the West.* New York: Penguin.

Watson, D. 1953. *Indians and the Mesa Verde.* Washington, DC: National Park Service.

West, P., and S. Brechin, eds. 1991. *Resident peoples and national parks: Social dilemmas and strategies in international conservation.* Tucson: University of Arizona Press.

PART VI

Searching for Solutions

Complexity. Conflict. Uncertainty. These are watchwords of the future if a warmer climate alters natural resources and human cultures of the Far North. These terms are generally used in a negative context. However, the problems associated with potential impacts of climate change may provide some opportunities as well.

The realization that humans can modify atmospheric and biological systems on a global scale has forced us to identify the critical scientific information needed to assess the current and future condition of natural resources. We can no longer afford to drift from one topic to the next in our scientific programs. Climate change may be imminent, and we need solid priorities for research, albeit within the constraints of shrinking budgets. An impending crisis has created an opportunity to develop a better organized approach for the study of natural resources.

Resource management needs to get its house in order too. As serious as current management problems may be, they pale in comparison to the specter of climate change. If our institutions and policies are incapable of resolving existing conflicts, how will they deal with natural and human systems under even greater stress? This is an opportunity to establish social processes and institutions that can deal with a dynamic biological and social environment.

Fortunately, some people are taking action. The Mackenzie Basin Impact Study demonstrates that it is possible to address interdisciplinary problems over a broad geographic area. Involving all of the stakeholders is the key. You must have participation in an issue to have ownership of the solution. Ongoing efforts in the Mackenzie Basin may well serve as a model for similar efforts in other parts of the world.

There is much work to be done if the biological and human systems of the Far North are to survive, with or without climate change. The collective recommendations of participants in the Human Ecology and Climate Change workshop—the original source of ideas for this book—are summarized briefly in the final chapter. An agenda of priority research and management actions articulates specific tasks that need to be implemented. This agenda issues a call for interdisciplinary approaches to all of our scientific and management problems. It emphasizes cooperation and communication. Above all, it emphasizes that we must act now.

The North is a special place for many reasons. We have a responsibility to keep it that way—for the caribou, the polar bear, the bowhead whale—and especially for the people who live there.

21

An interdisciplinary assessment of climate change on northern ecosystems: The Mackenzie Basin Impact Study

Stewart J. Cohen

Twenty years ago, the notion that humans could change the world's climate was considered only by scientists and environmentalists. Now, it is also a matter for governments. In March 1994, the Framework Convention on Climate Change (FCCC), one of the products of the 1992 Earth Summit, was ratified by the required fiftieth nation and became a part of international law. By August 1994, 80 nations had ratified it.

One could speculate on the various factors that have led more than half the world's nations to join in the battle against human-induced climate change, but for now, let us restrict our focus to the implications of this agreement. Much attention has been given to the "limitation" component, whereby nations are required to conduct emissions inventories and determine strategies for reducing emissions, or in the case of developing countries, for slowing the rate of growth of emissions. The ultimate objective, however, is to stabilize global concentrations of carbon dioxide and other greenhouse gases at a level that does not represent dangerous anthropogenic interference to the atmosphere. This effort hinges on the definition of the term *dangerous*.

This is an important challenge for climatic impact assessment. It is a complex multidisciplinary research challenge, but the ratification of the FCCC also makes it a challenge for decision makers in industry and government. To make matters even more difficult, it is an assessment not of an observed climatic event but of a theoretical warming of the earth's climate by increased concentrations of greenhouse gases (Chapter 2). This theory is supported by many scientists (IPCC 1990, 1992), although there are many uncertainties regarding the magnitude and

I thank Jenny Reycraft for producing Figure 21.1, Barrie Maxwell for his comments on this chapter and his guidance throughout MBIS, and the many participants in MBIS for their valuable contributions to northern research and for introducing me to the land north of 60°N. Any opinions expressed in this chapter are my own and not necessarily those of Environment Canada.

timing of changes (e.g., Lindzen 1990). If climatic warming occurs, the earth and its people will feel its effects and adapt to it. What might these effects be and what might (or should) be the nature of the adaptive responses in high latitude regions, particularly Canada's North?

Finding common ground

Global scale atmospheric events, such as the El Niño Southern Oscillation, affect different locations in unique ways (Glantz et al. 1987). Climate change may be a global challenge, but its impacts will be region-specific. This is also a political issue with implications for resource managers and regional economies. Scientific expertise is needed as well as regional knowledge gained through professional involvement with an economic, environmental, or social activity. No one discipline, agency, or interest group can provide the breadth required for such an assessment.

Some common ground is needed to attract this breadth of expertise and interests. Many impact assessments have focused on individual sectors (e.g., agriculture, wildlife, water resources), and while these provide technical information on "first-order" impacts (IPCC 1990, 1992; Tegart and Sheldon 1993), a wide range of external factors are often assumed to remain unchanged. Regional and national assessments are being produced elsewhere (e.g., Smith and Tirpak 1990; Hulme et al. 1992; Henderson-Sellers and Colls 1993), but these do not apply to Canada's North.

What should governments and society do about human-induced climate change? Limitation options (e.g., energy efficiency standards, carbon tax) are being considered on an international level, but there may still be a need for adaptation options (Smit 1993), especially because these can deal directly with region-specific impacts.

The interdisciplinary impact study of the Mackenzie River Basin described below is an attempt to produce an integrated regional assessment of scenarios of climatic warming for a large watershed in northern Canada. The regional focus has been used to attract scientific expertise and various stakeholders with local knowledge. The common ground for all of them is an interest in the future of this place.

The choice of study area can influence many aspects of an impact assessment, including issue identification and data collection. Various administrative and ecological settings might be considered, but the focus here is exclusively on watersheds as integrators. Land cover and land use affect hydrology and water quality, so water users (e.g., hydroelectric utilities, fisheries, navigation, wildlife, agriculture) are necessarily linked with forces that inadvertently or purposefully modify the landscape (e.g., agriculture, forestry, hydroelectric utilities, fire).

Governments have used watersheds as the basis for the creation of customized management structures (e.g., basin commissions, water boards). These attempt to reconcile the goals of competing interests while providing direction for regulation, water allocations, and other matters. For the purpose of climatic impact assessment, this is an important source of information on regional issues and their stakeholders. Watershed-based assessments have been previously applied to the Great Lakes (Smith and Tirpak 1990; Mortsch et al. 1993) and a series of international basin studies of the Nile, Indus, and Zambezi (Strzepek and Smith 1995).

Researching impacts in a multidisciplinary group setting

The Mackenzie Basin Impact Study (MBIS) is part of the Canadian government's Green Plan, and has now passed the halfway point in its six-year mandate to assess the potential regional implications of global climatic change (Cohen 1993, 1994). The MBIS program, which continues until 1997, includes studies on water resources, permafrost, vegetation, wildlife, economic activities, human communities, and applications of remote sensing and geographic information systems (GIS). Attention is also given to the challenges of producing an integrated assessment and to incorporating traditional ecological knowledge into MBIS.

This region was chosen because it is a major northern watershed (Figure 21.1) that has attracted considerable scientific attention in the past and also because of concerns about possible environmental impacts of resource development (e.g., MRBC 1981, NRBS 1994). Besides having the potential integrating features of a watershed, the Mackenzie Basin also contains many climate-sensitive landscapes and transition zones: tree lines (Arctic, montane, aspen parkland), discontinuous permafrost, wetlands and deltas, the edge of multiyear sea ice, and the northern limits of agriculture. Freshwater and terrestrial migratory wildlife might be sensitive to climate-induced changes in the landscape.

The Mackenzie is also a region of social transition. The southern third is a mid-latitude cash economy, much like the rest of North America. The Peace and Athabasca subbasins (Figure 21.2) contain agricultural land, and the southern boundary of the study area is within commuting distance of Edmonton, Alberta. The northern two-thirds is a mixture of single industry resource towns, government centers, and Native communities, some of which are not yet linked to all-season roads (e.g., Fort Norman, southeast of Norman Wells; Tuktoyaktuk, north of Inuvik). These communities follow a traditional lifestyle of wild food harvesting, which is a largely nonwage economy. Northern Canada is also undergoing a profound political transition with the creation of Nunavut and the current and impending agreements on Native land claims (see Chapter 11).

Figure 21.1. The Mackenzie Basin, Canada.

MBIS is attempting to produce an integrated regional assessment of global warming scenarios, as a way of identifying the indirect linkages between climate and regional policy concerns, such as land and water management. Several exercises are being tried, including (1) resource accounting with input-output modeling, (2) land assessment (including multiobjective program modeling), (3) review of water resources policy instruments, and (4) study of settlement patterns. Each of these utilizes the outputs of various individual studies to address some of the human dimensions of climate change. Traditional knowledge studies may contribute to integration because they could provide important "ground truth" for broader studies. There is no consensus on which method is best for producing an integrated study: multiple approaches are being used as a "family" of integrators that address complementary issues. The interdisciplinary group approach of MBIS, which combines scientific inquiry with regional stakeholder consultation, is also an important integrating function. A successful integration exercise will be an important advance in the development of impact research methodology.

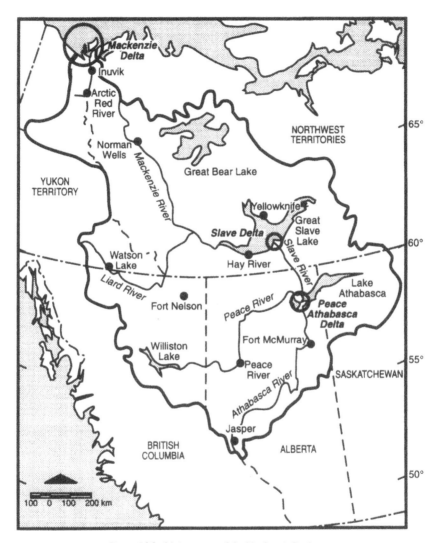

Figure 21.2. Major rivers of the Mackenzie Basin.

The horizontal integration challenge

Several aspects of "integration" relate to the process of linking scientific disciplines to each other. Examples of horizontal integration include information exchange between hydrology and wetland ecology; wildlife and Native studies; or forestry, fire, permafrost and tourism. This might seem like a daunting task. Scientists tend to specialize, although interdisciplinary research is receiving increased recognition and support. New journals (such as *Global Environmental Change*), new research institutes and funding arrangements (e.g., the Sustainable Development Research Institute at the University of British Columbia and Canada's Tri-Council fund), and even the production of this book are encouraging signs. Researchers and stakeholders come from backgrounds that provide each of them with unique perspectives. How can these be combined so that the whole picture is represented?

There are several opportunities to facilitate this linkage in MBIS. Integrated system models, economic models, and other similar tools can be used to address complex issues related to land use and economic growth (e.g., Lonergan 1994; Yin and Cohen 1993,1994; Huang et al. 1994). These mathematical or statistical techniques require a wide range of inputs, including census data, outputs of other models, and/or indices obtained from remote sensing, thereby serving as integrators of information obtained from other disciplines.

Management and legal instruments (e.g., economic development plans, water management agreements, land use plans) may also respond to first-order impacts on the landscape and represent the integration of scientific information and stakeholders' preferences. Their performance under climate change scenarios would provide an important measure of impact. Policy analysis has both quantitative and qualitative aspects and may be preferred by stakeholders who are leery of "black box" models.

Scales, units, and other language barriers

Horizontal integration must consider the challenge of reconciling differences in language, data, and perception. An example within MBIS is interjurisdictional differences in data collection networks (e.g., forest districts, crop reporting districts, counties, subbasins, land claim areas, climatic stations). Each operates on its own spatial and temporal scale, and they all differ in their representativeness. This problem is exacerbated in northern Canada by administrative differences between provinces and territories, and low population densities with sparse monitoring networks.

There are some options for creating a regional database of spatial information through the use of remote sensing and GIS. The reliability of these data sets depends on the availability of ground truth; expert judgment may be the only available source for some areas. This is a challenge, but also an opportunity, for merging modern methods with traditional knowledge.

Collaboration with stakeholders

After completing preliminary consultations in 1990, MBIS established a Working Committee, composed of representatives from government agencies and nongovernment stakeholders. Membership was drawn from those involved in preliminary consultations, but the Working Committee expanded to 30 members as other interested parties became aware of the effort. Within the Working Committee, an Integration Subcommittee is composed of physical, biological, and social scientists and regional stakeholders. This subcommittee has only 8–12 members to facilitate discussion on study issues; principal investigators developing integration targets (e.g., resource accounting) are included. MBIS is considering regional concerns in study design through this process. The intent is to provide stakeholders with a sense of ownership and increase the likelihood of regional acceptance of study findings and recommendations.

Vertical integration challenge

The linkage of science to policy, or vertical integration, has been evolving during the planning, consultation, and research phases of this program. Participation of senior government managers who can articulate the information requirements of policymakers, has been critical. At a workshop held in February 1992, representatives from several Alberta, Northwest Territories, and Canadian government agencies discussed issues with MBIS participants. This exercise identified five adaptation policy targets (Cohen 1993): (1) interjurisdictional water management, (2) sustainability of Native lifestyles, (3) economic development opportunities, (4) maintenance of infrastructure, and (5) sustainability of ecosystems. The breadth of these policy targets indicates the scope of the vertical integration challenge. Within the time and resource constraints of MBIS, research activities need to be initiated and completed with these targets in mind. When a research proposal is reviewed by the Working Committee, both good science and relevance to policy are considered.

Ivory towers and Northern realities

While the vast, sparsely populated North is rapidly changing, the northern research agenda is also changing. The creation of Nunavut, the completed Inuvialuit and Gwich'in claims, and the soon-to-be settled Sahtu claim have produced a change in the marketplace for information. As Native communities take on greater responsibilities for resource management, they will also play a greater role in research and policy development. This includes control of scientific research licenses, intellectual property rights covering traditional ecological knowledge, and direct participation in wildlife and community studies. They want to know if proposed research will be relevant to their needs, and they want information communicated directly to them, orally if necessary. The Inuvialuit, for example, have already produced Community Conservation Plans that outline their interests in resource management and research (e.g., Inuvik 1993). Publications in scientific journals alone are not sufficient.

Can climatic impact assessments like MBIS, filled with long-term scenarios and uncertainties, be effectively communicated to Northerners? Analogies may be more useful than abstract descriptions of model results (Glantz 1988; Kearney 1994), but it is difficult to identify an appropriate analogy for a warmer North, other than the time when the Norse colonized Greenland. The next century's challenges will be considerably different from those of 1,000 years ago, and climate change will have to be considered alongside many other concerns.

Climate change and merging agendas

Short-term and long-term issues

With land claims and other political changes attracting so much immediate attention, it is understandable that a long-term issue such as climate change can be ignored or not taken seriously. The global warming debate seems far away to the residents of Aklavik, although residents do notice changes in the local environment that may be related to climate (Aharonian 1994).

Despite the focus on more immediate concerns, there is genuine interest in environmental issues and growing awareness of climate change among resource managers and administrators for the new Native organizations. The MBIS Working Committee currently includes representatives from the Inuvialuit Game Council, Gwich'in Tribal Council, Dene Nation, and Métis Association of the Northwest Territories. There have been consultations with staff from the Sahtu and Deh Cho Tribal Councils and the communities of Lutsel k'e, Aklavik, and Fort Liard.

Canada's Northerners have also experienced the boom and bust of nonrenewable resource development and are facing the prospect of new activities in diamond mining and oil and gas production. This kind of economic activity is a mixed blessing because most of the benefits go to non-Northerners. However, these initiatives do bring jobs and opportunities for Northerners to participate in the wage economy, which can supplement Native communities' hunting and fishing activities. These traditional activities enable Natives to meet most of their food requirements at a much lower monetary cost than reliance on imported food.

This juxtaposition of wage and nonwage economies is an important social issue in the North and needs to be accounted for in research efforts like MBIS. The equivalent monetary value of the nonwage economy is not well-documented, but must be included to determine social and economic impacts of climate change that are relevant to Northerners. Assessing nonwage economies and the future of Native communities in a climate change context inevitably leads to concerns about the policy implications of the study itself. One community had originally expressed opposition to MBIS because of fears that a study-based scenario result could lead to government using climate change as an excuse to reduce social services. The community had to be reassured that the study would provide a basis for discussion about policy among all stakeholders, not just higher levels of government, and that it is an "arm's length" scientific effort.

MBIS is not only an exercise in interdisciplinary research with stakeholder collaboration, but also mediates the debate over regional implications of climate change. Because there are many other issues related to regional futures, MBIS is inadvertently becoming a forum for other agendas.

Climate change as a means to other ends?

A growing list of issues in discussions about MBIS includes conservation, traditional knowledge, nonrenewable resource development, Native self-government, Arctic science, Natives as research subjects, Natives as researchers, and local control of Arctic science. All of these agendas meet in MBIS with unanticipated results.

Discussions about the regional implications of climate change scenarios can set the stage for other issues as well. Examples include wage employment for Natives, requests for immediate intervention in the drafting of legislation, and fears of what the energy industry might say. If the industry disagrees on the need for controlling emissions, how could it support studies on the impacts of possible scenarios that might strengthen the case for stronger emission controls? Meanwhile, those who favor stronger controls do not want to consider potential adaptation responses to climatic impacts because it might weaken the case for controls.

Although MBIS is a research effort, the consultation process has had its moments of tension. While some have suggested that shortcomings are deliberate acts, there is no easy way to consult with all interested parties. Long-term climate change is commonly perceived as less important than short-term concerns, making it difficult for stakeholders to justify investing time and human resources. As a result, they may not be able to accept invitations to participate in study meetings and workshops, and they may be unfamiliar with recent scientific opinion.

There are additional constraints in Canada's North. Few researchers live there, so travel costs for meetings held in the study area are high. Northern-based Native organizations include support staff who focus on environmental issues, but they have to deal with immediate concerns and do not have much time to devote to climate change. Because staff size is usually small, staff changes can affect the continuity of consultation with MBIS. In addition, the land claims process has led to changes in administrative structures, as broad umbrella networks (e.g., Dene Nation) are now being complemented by programs of individual Tribal Councils. This rapid evolution has created a moving target for consultation, and as the North's political future evolves, the management of scientific research programs in the North will have to evolve with it. These additional contacts may provide more opportunities to enhance consultation and dialogue between scientists from North and South.

A *midterm progress report*

Climate as an agent of change

One theme that has clearly emerged in MBIS is that climate is a complex agent of change. Although scientific and political discussions have tended to focus on atmospheric change (Chapter 2), the land and its people will likely experience climatic warming through changes in stream flow, water levels, ice and snow cover, permafrost, plant growth, wildlife patterns, fire, pests, and diseases (Chapters 5 and 6). Some changes may occur gradually while others may come in the form of large steps or new extremes.

The linkage between changes in air temperature and regional socioeconomic concerns is largely through these landscape filters. Biophysical changes are what people will notice before they pay attention to climatic statistics. Has the winter road season changed? Is anything new with the caribou migration? Are current fire management strategies still working satisfactorily? What is the status of permafrost along the Mackenzie Valley and the Beaufort coastal zone?

Some preliminary indications for the scenarios being assessed by MBIS are (1) spring flood risks may not be very different from current conditions (Newton 1994, Soulis et al. 1994); (2) fire hazard may become more severe (Kadonaga 1994), with subsequent effects on methane production and changes in local vegetation from spruce and wetland species to a post-fire mixture of deciduous trees, grassland, and weedy species (Wein et al. 1994); and (3) permafrost thaw could lead to serious erosion problems in certain areas (Egginton 1993; Solomon 1994) but may be delayed elsewhere (Vitt and Halsey 1994).

These first- and second-order questions eventually lead to others that are considerably more difficult to address. Will land claims or water resources agreements be affected? Could there be new conflicts over land use? What might be the effects on parks and other protected areas? Could climate change affect the economics of oil and gas production in the Beaufort Sea?

Can MBIS achieve integration?

The answers to higher-order questions require methods that combine information from various disciplines. The result is more than the sum of the parts. Integrated regional assessment is a lofty but somewhat fuzzy objective, and there are few successful examples. Partial integration means that there are some aspects that remain outside the assessment, and assumptions have to be made about their level of influence. For example, there are no research activities that focus on marine life in the Beaufort Sea, Native communities in Alberta, or future economic linkages with the rest of Canada. MBIS includes population and economic growth scenarios, but technological and institutional change scenarios have not been constructed. Are these shortcomings? Probably. Does this prevent MBIS from achieving full integration? Yes. Does this prevent MBIS from providing relevant information on regional impacts? Hopefully, no.

If MBIS can address the major questions previously discussed with the acceptance of both the scientific community and regional stakeholders, one could argue that a sufficient level of integration has been achieved. Scientists will express their judgment through the review process associated with publication of results in the scientific literature. Stakeholder opinion may come through various newsletters, statements to the media, or other forums. Consensus about all issues is unrealistic, but general acceptance of MBIS as a serious effort worthy of consideration by the region's stakeholders would be a positive sign.

Can MBIS achieve stakeholder acceptance?

Scientists are generally not confronted by a review committee of stakeholders, but future collaborations with stakeholders on climate change will be influenced by previous experience. MBIS is acting as a broker between science and policy, while acting in an arm's length capacity to produce information for a broad range of public interests. Stakeholder acceptance will likely depend on whether or not participants in MBIS are perceived in that way. MBIS is an initiative of the Canadian government, not the region itself. This is not new. Interjurisdictional cooperation in the North is not new either. However, there is a feeling (based on previous experience) that federally driven research may not be in the best interests of Northerners, or that cooperation is sought merely to get others to contribute funding without having input on the research.

Funding for this exercise would have been insufficient without the Canadian government's Green Plan. However, Northerners often feel that they are left out of the southern-based information network. More Northerners need to be involved in these efforts, but that will not happen unless there is something in it for them. Stakeholders have been brought into MBIS from the beginning through the Working Committee and various MBIS meetings or workshops. If they feel a sense of ownership in the project, perhaps that will carry through to the end, when results are presented and responses considered.

Life after MBIS?

The 1994–97 period will include completion of the first-order biophysical impact studies, transfer of information to the integrators (systems modelers, policy analysts, etc.), completion of integration exercises, publication of the final report, and communication of the results to regional stakeholders and in the scientific literature. Information transfer and communication activities are the key element. These will be challenging, and there are no guarantees that all information will get through the various filters without mistakes in translation.

There will be another workshop similar to the event that facilitated the production of *MBIS Interim Report #2* (Cohen 1994). MBIS investigators are expected to exchange information with each other before and after their components are completed. There are also plans for more discussions on MBIS within the region, before and after publication of the final report in 1997. These will take place in various communities with the assistance of the Science Institute of the Northwest Territories and regional stakeholders. If communication is exclusively through newsletters, journal articles, and other written forms, then information from the project will not reach Northerners.

Lessons for other impact assessments

It may be difficult at this stage to appreciate the long-term value of the MBIS experience. However, one thing is clear: Collaboration with stakeholders is vital for producing an assessment that is useful and relevant to the region of interest. In fact, partially or fully integrated assessments may be impossible without stakeholder involvement during all phases of research.

The climate change issue has its scientific and political sides, and this will not change. Both communities need to be engaged from the beginning of any impact assessment. If the science and science management are executed competently, the assessment should withstand a credible review process. But there needs to be some assurance that the study committees are truly representative of scientific and stakeholder interests, and that the parties agree on the assessment's mission before the exercise is launched.

MBIS has attracted a modest amount of interest from Northerners and other stakeholders, but has difficulty competing with short-term agendas needing urgent attention. The fact that not everyone is concerned about global warming is related to the narrowness of the public debate at the national and local levels. One example is the current focus on emission controls without considering obstacles to adaptation (Smit 1993). Another is the lack of an unambiguous "smoking gun." Some have argued that international opinion was spurred on by the drought and heat wave that struck the North American midcontinent in 1988, only to be dissipated by the cooling of 1992 brought on by the eruption of Mount Pinatubo. What kind of "smoking gun" are Northerners looking for? Widespread landslides or slope failures brought on by permafrost thaw? Extensive forest fires in the boreal zone? A noticeable reduction in the ice season? A change in caribou migration patterns? If the "smoking gun" cannot be detected at some point, many will become convinced that it does not exist and that the time has come to move on to other issues. Trust, good will, and scientific curiosity will not be enough to hold everyone together.

Summary and a final note: It isn't over yet

The MBIS exercise has already produced tangible results: (1) new participants in a growing multidisciplinary northern research network, (2) increased regional awareness of a complex externally driven environmental issue, and (3) increased awareness among (a few) Southerners about the North. If MBIS can convince

Northerners to acquire some ownership of the climate change research agenda for their lands, they may also put a northern stamp on responses.

Since 1990, MBIS has traveled on a bumpy road through uncharted territory. It has its flaws, but like all pioneering experiments, it has generated a considerable amount of interest from within Canada and from other countries. We now know a great deal more about the potential impacts of climate change in the North. Furthermore, the experiences of MBIS will benefit future multidisciplinary assessments of climate change and other environmental impacts.

References

Aharonian, D. 1994. Climate-society interactions in Aklavik, NWT. In *Mackenzie Basin Impact Study (MBIS) interim report #2*, edited by S. J. Cohen, 410–420. Downsview, Ontario: Environment Canada.

Cohen, S. J., ed. 1993. *Mackenzie Basin Impact Study interim report #1*. Downsview, Ontario: Environment Canada.

Cohen, S. J., ed. 1994. *Mackenzie Basin Impact Study (MBIS) interim report #2*. Downsview, Ontario: Environment Canada.

Egginton, P. 1993. Permafrost south of the Beaufort coastal zone. In *Mackenzie Basin Impact Study (MBIS) interim report #1*, edited by S. J. Cohen, 52–58. Downsview, Ontario: Environment Canada.

Glantz, M. H., ed. 1988. *Societal responses to regional climatic change: Forecasting by analogy*. Boulder, CO: Westview Press.

Glantz, M. H., R. Katz, and M. Krenz, eds. 1987. *The societal impacts associated with the 1982–83 worldwide climate anomalies*. Boulder, CO: National Center for Atmospheric Research.

Henderson-Sellers, A., and K. Colls, eds. 1993. Climatic impacts in Australia. *Climatic Change* 25(3–4, special issue):201–438.

Huang, G. H., Y. Y. Yin, S. J. Cohen, and B. Bass. 1994. Interval parameter modelling to generate alternatives: A software for environmental decision-making under uncertainty. In *Computer techniques in environmental studies*, edited by C. A. Brebbia, 213–223. Ashurst Lodge, UK: Computational Mechanics Publications.

Hulme, M., T. Wigley, T. Jiang, Z.-c Zhao, F. Wang, Y. Ding, R. Leemans, and A. Markham. 1992. *Climate change due to the greenhouse effect and its implications for China*. Gland, Switzerland: World Wildlife Fund for Nature.

Intergovernmental Panel on Climate Change (IPCC). 1990. *Climate Change: The IPCC scientific assessment*, edited by J. T. Houghton, G. J. Jenkins, and J. J. Ephraums. Cambridge: Cambridge University Press.

Intergovernmental Panel on Climate Change (IPCC). 1992. *Climate change 1992: The supplementary report to the IPCC scientific assessment*, edited by J. T. Houghton, B. A. Callander, and S. K. Varney. Cambridge: Cambridge University Press.

Inuvik, Community of. 1993. *Inuvik Inuvialuit community conservation plan*. Inuvik, N.W.T.: Wildlife Management Advisory Council.

Kadonaga, L. 1994. Fire in the environment. In *Mackenzie Basin Impact Study (MBIS) interim report #2*, edited by S. J. Cohen, 329–336. Downsview, Ontario: Environment Canada.

Kearney, A. R. 1994. Understanding global change: A cognitive perspective on communicating through stories. *Climatic Change* 27:419–441.

Lindzen, R. S. 1990. Some coolness concerning global warming. *Bulletin of the American Meteorological Society* 71:288–299.

Lonergan, S. 1994. Natural resource/environmental accounting in the Mackenzie Basin. In *Mackenzie Basin Impact Study (MBIS) interim report #2*, edited by S. J. Cohen, 39–42. Downsview, Ontario: Environment Canada.

Mackenzie River Basin Committee (MRBC). 1981. *Mackenzie River Basin study report*. Regina, Saskatchewan: Environment Canada.

Mortsch, L., G. Koshida, and D. Tavares, eds. 1993. *Adapting to the impacts of climate change and variability: Proceedings of the Great Lakes-St. Lawrence Basin Project workshop*. Downsview, Ontario: Environment Canada.

Newton, J. 1994. Community response to episodes of flooding in the Mackenzie Basin. In *Mackenzie Basin Impact Study (MBIS) interim report #2*, edited by S. J. Cohen, 421–430. Downsview, Ontario: Environment Canada.

Northern River Basins Study (NRBS). 1994. *Status report*. Edmonton: Northern River Basins Study.

Smit, B., ed. 1993. *Adaptation to climatic variability and change, report of the Task Force on Climate Adaptation, Canadian Climate Program*, Department of Geography Occasional Paper 19. Guelph, Ontario: University of Guelph.

Smith, J. B., and D. A. Tirpak, eds. 1990. *The potential effects of global climate change on the United States*. Report to Congress. Washington, DC: United States Environmental Protection Agency.

Solomon, S. 1994. Storminess and coastal erosion at Tuktoyaktuk. In *Mackenzie Basin Impact Study (MBIS) interim report #2*, edited by S. J. Cohen. Downsview, Ontario: Environment Canada.

Soulis, R., S. A. Solomon, M. Lee, and N. Kouwen. 1994. Changes to the distribution of monthly and annual runoff in the Mackenzie Basin under climate change using a modified square grid approach. In *Mackenzie Basin Impact Study (MBIS) interim report #2*, edited by S. J. Cohen, 197–209. Downsview, Ontario: Environment Canada.

Strzepek, K. M., and J. B. Smith, eds. 1995. *As climate changes: International impacts and implications*. New York: Cambridge University Press.

Tegart, W. J. M., and G. W. Sheldon, eds. 1993. *Climate change 1992: The supplementary report to the IPCC impacts assessment*. Canberra: Australian Government Publishing Service.

Vitt, D. H., and L. A. Halsey. 1994. Disequilibrium response of permafrost in peatlands of the southern Mackenzie Basin. In *Mackenzie Basin Impact Study (MBIS) interim report #2*, edited by S. J. Cohen, 275–277. Downsview, Ontario: Environment Canada.

Wein, R., R. Gal, J. C. Hogenbirk, E. H. Hogg, S. M. Landhäusser, P. Lange, S. K. Olsen, A. G. Schwarz, and R. A. Wright. 1994. Analogues of climate change-fire-vegetation responses in the Mackenzie Basin. In *Mackenzie Basin*

Impact Study (MBIS) interim report #2, edited by S.J. Cohen, 337–344. Downsview, Ontario: Environment Canada.

Yin, Y., and S. J. Cohen. 1993. Integrated land assessment framework. In *Mackenzie Basin Impact Study (MBIS) interim report #1*, edited by S. J. Cohen, 151–163. Downsview, Ontario: Environment Canada.

Yin, Y., and S. J. Cohen. 1994. Identifying regional policy concerns associated with global climate change. *Global Environmental Change* 4:246–260.

22

An action plan for an uncertain future in the Far North

David L. Peterson and Darryll R. Johnson

There has always been variation in the biological and human systems of the Far North. Past climatic changes have altered the distribution and abundance of natural resources, and human cultures have responded with various behaviors and adaptations. Despite the resilience of natural and human systems, it is clear that they will be at considerable risk if subjected to a rapidly changing climate.

There is, of course, a great deal of uncertainty about the future. Until compelling evidence of climate change is detected, we can make only educated guesses about the magnitude and rate of changes in the atmospheric, biological, and social environment. However, the potential implications of these changes are so great that we must take action before the changes are upon us. Western society has created this potential crisis in the North. It has a responsibility to anticipate future changes and act assertively to address them.

How should we act in the face of such uncertainty? First, we need a better understanding of northern ecosystems to identify the potential impacts of climate change on natural resources. Second, we need to ensure that responsible human institutions, which can address human concerns and needs in a changing environment, are in place. The authors of this book and other participants in the Human Ecology and Climate Change workshop held in October 1993 considered these problems and articulated a number of issues that need to be addressed during the next decade.

These issues are categorized as scientific or management to provide some structure to the discussion and proposed actions. However, it must be strongly emphasized that we need to look for interdisciplinary solutions wherever possible. The lines between research and management, and between biological science and social science, should become increasingly blurred to facilitate communication and cooperation at all levels.

Priorities for the future

A basic understanding of the impacts of past climate change is critical to understanding potential effects in the future. Considerable effort has been devoted to reconstruction of paleoclimates and paleoecological conditions. But we also need to more fully investigate the role of climate on cultural aspects of human populations. The past can be visualized as a laboratory in which various experiments and their outcomes are recorded. Climatologists, biologists, and social historians should be working together to learn more about that laboratory. If the past is examined from an interdisciplinary perspective of process, rather than for specific artifacts, it can be a fertile source of information for forecasting and model-building.

Scientific issues and actions

Extensive inventories and monitoring of natural resources and human populations are needed now—before any dramatic changes occur—to establish a baseline for future evaluations. Specific and general predictive models should also be developed, but these should have a lower priority for the time being. Empirical data can be incorporated into these models as they become available. Priorities for a better understanding of the physical and biological environment include the following:

- Study tension/transition zones and physiological limits of organisms, where changes should be most observable (e.g., latitudinal tree line)
- Identify sources and sinks of carbon storage over a broad range of northern environments; include short-term manipulations (e.g., soil heating) and long-term monitoring
- Link animal population responses to changes in vegetation, hydrology, and the marine environment; focus on wildlife movements in protected areas
- Determine how indirect effects of climate change (e.g., fire frequency, decomposition, permafrost changes) would amplify direct effects and feedbacks
- Study the effects of climate on the lower end of the food chain in marine environments (e.g., the effects of upwelling on productivity and prey organisms)
- Study the long-term impacts of climate on ice and hydrologic cycles
- Develop better predictive models with respect to regional climate and biophysical impacts, with emphasis on surface energy budgets
- Describe the relationship of subsistence activities to the distribution and abundance of resources, especially in protected areas, and determine how subsistence-wildlife relationships might be altered by climate change

Priorities for a better understanding of sociocultural systems include:

- Establishing studies on the patterns and types of existing conflicts between specific communities and institutions as a baseline for evaluating future changes
- Incorporating an *emic* approach in all research designs (This approach considers values relevant to human culture and understanding, and should include the knowledge of elders who have observed and "managed" natural resources throughout their lives.)
- Determining which social/managerial institutions most effectively integrate local populations in decision making and would be most beneficial under changing environmental and social conditions
- Determining how local people perceive potential changes in climate and other environmental characteristics
- Characterizing current patterns of subsistence activities (changing from generation to generation) and identify social constraints (e.g., legal, territorial) on subsistence harvest
- Developing projections of human population growth and resource use to provide better predictions of future climate change impacts
- Determining how new (scientific) information can be integrated into resource management policy and rural culture
- Quantifying how changes in the material economy affect social systems

Studies of the northern biophysical environment should not be isolated from studies of social issues and local communities. Involvement of local people in research and other scientific activities should be encouraged. Furthermore, scientists have the responsibility of communicating their results effectively to local resource managers and communities. The value of traditional ecological knowledge and cultural perspectives should be included to a greater extent in data collection and interpretation.

Managerial issues and actions

Managers of protected areas and other lands in the Far North have a nearly impossible task, even without the challenge of a rapidly changing climate. Most resource managers are faced with complex environmental and social issues in large geographic areas, but have insufficient staff and budget to address these issues. Setting priorities is therefore critical. Furthermore, managers who are trained in "the South" and are charged with administering policies developed in Washington, D.C., or Ottawa soon learn that the rules are often different in the North. Sociocultural norms and world views are different, and problem solving requires patience, understanding, and creativity. Rigid policies and failure to include local people in managerial decisions can result in a hostile political environment and ineffective management.

The potential impacts of climate change might seem academic to resource management staffs who are already overburdened with responsibilities. Resource managers tend to respond cautiously to uncertain conditions in any case. The most likely action is often to maintain the status quo. Rather than focus on the specific impacts of climate change, it may be more effective to develop managerial priorities that address all types of environmental and social changes at the local level. Priorities for effective management include:

- Preparing management plans that respond to various types of potential environmental change (This should be done cooperatively with all appropriate agencies and stakeholders.)
- Establishing a policy goal of expanding the range of acceptable resource management practices; building flexibility into management plans and policies
- Developing comanagement arrangements among all relevant landowners and managers (local, regional, federal, Native, non-Native, etc.)
- Developing a cross-cultural planning process that involves local people and all stakeholders from beginning to end
- Promoting the concept of ecosystem management, including both biophysical and social issues
- Developing partnerships among resource management agencies at the local and regional level
- Developing mechanisms for sharing data and communicating at local, regional, and national levels
- Emphasizing and developing educational programs and communication at the local level
- Working for additional legislative and funding support for resource management problems in the North, with emphasis on the potential for rapid environmental change
- Assigning the best and brightest leaders to manage Northern protected areas and other public lands

In addition to these general approaches to management and institution-building, it is imperative that climate change becomes a discrete item on the management agenda. Resource managers will increasingly need a strong scientific background to understand climate change and other complex environmental problems. There must be a better link between management and research in the future, so that data collection and analyses are relevant to the needs of land managers.

Small steps and giant leaps

The issues and actions cited above are not intended to be comprehensive or to provide the ultimate road map for dealing with natural resource and sociocultural issues in the Far North. They are only the first steps in addressing existing problems and potentially more critical future problems. Taking those first steps must be a priority.

Alaska and the Yukon and Northwest Territories comprise 544 million hectares—600 Yellowstone National Parks, in terms of land area. It requires a leap of imagination to consider the magnitude of the impacts of climate change in this region. It also requires some new and creative ways of thinking about natural resources and human populations. The most sophisticated general circulation models barely include snow cover in their predictions of future climates. Inuit have 25 descriptive words for snow (Table 22.1; Dorais 1990). Western scientists have identified hundreds of plant species in the Far North. Inuit make no distinction between flowers, grass, or any other plants unless they are specifically used for something. There are sharply contrasting world views here. There are also some perspectives that can be shared between disciplines and cultures.

In Chapter 9 of this volume, we are reminded that not every response to environmental change in the Far North has been successful. Northeastern Greenland was the last area reached by the Thule people (ca. 1500 A.D.), who rapidly spread the technology of hunting bowhead whales (*Balaena mysticetus*). Archeological data indicate that as climate cooled over the next few hundred years, they aban-

Table 22.1. *Various types of snow, as expressed in Arctic Quebec Inuktitut.*

Inuktitut	English
qanik	falling snow
qanittaq	recently fallen snow
aputi	snow on the ground
maujaq	soft snow on the ground
masak	wet falling snow
matsaaq	half-melted snow on the ground
aqilluqaaq	drift of soft snow
sitilluqaq	drift of hard snow
qirsuqaaq	refrozen snow
kavirisirlaq	snow rendered rough by rain and freezing
pukak	crystalline snow on the ground
minguliq	fine coat of powdered snow
natiruvaaq	fine snow carried by the wind
piirturiniq	thin coat of soft snow deposited on an object
qiqumaaq	snow whose surface is frozen
katakartanaq	hard crust of snow giving way under footsteps
aumannaq	snow ready to melt, on the ground
aniu	snow for making water
sirmiq	melting snow used as cement for the snowhouse
illusaq	snow that can be used for building a snowhouse
isiriartaq	yellow or reddish falling snow
kinirtaq	damp, compact snow
mannguq	melting snow
qanialaaq	light falling snow
qanniapaluk	very light falling snow, in still air

Source: Dorais (1990), p. 205.

Index

Note: t indicates table and f indicates figure.

Wind patterns, global, 18–21, 20f
Wind power, 32
Wolves, 225
  control programs, 296
  red wolf (*Canis niger*), 270
  symbolic significance of, 296

Yukon
  age pyramids, 38f
  population projections, 45
Yukon Kuskokwim Wildlife Refuge, 292
Yukon River flooding, 150–151

Yukon Umbrella Final Agreement, 199
Yup'ik
  epistemological assumptions, 223
  oral tradition, 150–151
  subsistence patterns, 141
  traditional knowledge of species habits,
    225
  (*See also* Eskimo, Inuit, Inupiat)

Zonation models for biosphere reserves,
  262–265, 263f, 264f